Premature Bonanza

Premature

Bonanza

Standoff at Voisey's Bay

Mick Lowe

Between the Lines

Premature Bonanza: Standoff at Voisey's Bay
© Mick Lowe, 1998

All rights reserved. No part of this publication may be photocopied, reproduced, stored in a retrieval system, or transmitted in any form or by any means, electronic, mechanical, recording, or otherwise, without the written permission of Between the Lines, or CANCOPY (photocopying only), 6 Adelaide Street East, Suite 900, Toronto, Ontario, M5C 1H6.

Between the Lines gratefully acknowledges financial assistance for our publishing activities from the Ontario Arts Council, The Canada Council for the Arts, and the Government of Canada through the Book Publishing Industry Development Program.

Parts of this book were first published, in a different form, in *Canadian Forum* and *Elm Street* magazines, and in *Northern Life*, Sudbury's community newspaper.

Every reasonable effort has been made to find copyright holders. The publisher would be pleased to have any errors or omissions brought to its attention.

Canadian Cataloguing in Publication Data

Lowe, Mick, 1947-
    Premature bonanza : standoff at Voisey's Bay

Includes bibliographical references and index.
ISBN 1-896357-17-2

1. Nickel mines and mining – Newfoundland – Voisey Bay.
2. Inco Limited.   I. Title.

HD9539.N52V65 1998a   338.7'6223485'097182   C98-932072-3

Cover Photo and Interior Photos (Pages 101-105): Ted Ostrowski
Interior Photos (Pages 106-108): Mick Lowe
Interior Design: Gordon Robertson

Printed in Canada by Transcontinental

1 2 3 4 5 6 7 8 9 10   05 04 03 02 01 00 99 98

Between the Lines
720 Bathurst Street, #404, Toronto, Ontario, M5S 2R4, Canada
(416) 535-9914   btlbooks@web.net

for Anne-Marie
muse, grounding rod, balance wheel

ALSO BY MICK LOWE

Books

> *Conspiracy of Brothers: A True Story of Bikers, Murder and the Law*, (1988)
>
> *One Woman Army: The Life and Times of Claire Culhane*, (1992)

Radio Play

> *The Organizer's Tale*, (1982)

Contents

Prologue   1

Enter Inco *June 1995–July 1996*

1 Duelling by Degrees   7
2 Rosencrantz and Guildenstern 'r Us   25

A Trip to Voisey's Bay *October–November 1996*

3 Welcome to Nain   33
4 Voisey's Bay at Last   39
5 Who Owns Voisey's Bay?   45
6 The View from St. John's   51
7 Irresistible Force, Immovable Object   57

A Season of Discontent *Summer 1997*

8 Inco Drills a Bootleg   67
9 Seven Days at Voisey's Bay   77
10 The Bootleg Explodes   89

Photo Section   101–108

Voisey's as Vortex *Fall–Winter 1997–98*

11 Nickel & New Caledonia:
    They Shoot Kanaks, Don't They?   111
12 Sui Generis (1)   119
13 Sui Generis (2)   133
14 Spooning Pablum to the Bay Street Gang   139
15 Can Inco Be Trusted?   147

Voisey's Redux *February 1998*

16 No Smelter, No Mine?   157
17 "We Are the Innu of Northern Labrador..."   161
18 Disarmed and Dangerous:
    Katie Rich on Home Ground   169
19 Notes from the North Coast   179

Epilogue *August 1998*

20 Exit Inco?   189

Afterword and Acknowledgments   197
Appendix: Joan Kuyek's Commencement Speech   199
Annotated Bibliography   207
Glossary   215
Index   219

"We walk on the land. The Akeneshau (white man) is different. His feet never touch the ground. He walks on pavement, and his feet are off the ground all the time. Then he comes to Labrador and says, 'What a beautiful country!'"

– Akat Piwas, Innu elder
Between a Rock and a Hard Place

Prologue

WHAT A BEAUTIFUL COUNTRY! To the people who live there and whose feet have always touched the ground, Voisey's Bay, it turns out, is a kind of sub-Arctic oasis. The Innu and Inuit of Labrador have long prized the place because it is reasonably sheltered from the worst of the incessant winds that scour the North Labrador coast, because it is well-treed, and because the three streams ("brooks," they are called here) that braid the land are rich in trout and Arctic char. For all of these reasons black bear are common, and sometimes caribou, strays from the great George River herd, pass through on their annual migrations. To the Innu and Inuit, then, this place has always been like money in the bank: a stable, secure source of sanctuary, timber, fresh water, and food.

Picture a left hand, palm down, fingers extended as if grasping for Ireland, thousands of kilometres to the east across the icy North Atlantic. The fingers are part of the crenellated coastline of Northern Labrador. On the back of the hand is a high, windswept hill overlooking the Labrador Sea, and sizeable bays lie both above the hand, to the north, and below it, to the south.

PREMATURE BONANZA

The people of the Innu Nation called the bay to the south Emish, an Inuit derivation of "Amos," because a white trader, Amos Voisey, had set up shop there in the 1930's. Emish, or Voisey's Bay, was the northerly limit of Nitassinan, the vast Innu homeland that has stretched for millennia from the east coast of south central Labrador far inland to what is now the Province of Quebec.

To the Inuit the bay to the north was Anaktalak, literally "Place of Much Caribou Shit." Anaktalak was the southerly limit of this largely coastal people, who were themselves among the southernmost members of the earth's original circumpolar inhabitants.

The land lay as it always had—majestic, pristine—and this day burnished by a late fall sun, when a helicopter flew over the rocky headland between Emish and Anaktalak late in the afternoon of a day in September 1993. Aboard, staring moodily at the ground below, were two obscure Newfoundland prospectors named Al Chislett and Chris Verbiski. They had begun the exploration season in the spring with a fat bankroll and high expectations. Their company, Archean Resources Ltd., had entered into a joint venture with Diamond Fields Resources Inc. of Vancouver to search for diamond deposits in the forbidding terrain of Northern Labrador. The pair had started with $450,000 and a full exploration crew. Now, four months later, after exploring 25,000 square kilometres of land, they found their money gone. They also had found no diamonds. Chislett and Verbiski laid off their employees and took off for Nain, defeated. But as they passed over the bald hill between two bays south of Nain, they noticed a peculiar reddish colour in the exposed rock.

Such colouration, they knew, is often a sign of oxidation—rust—the product of weathering on exposed rock with high mineral content. They made a mental note to return, and when they did they chipped away some of the reddish rock, "grab samples," in the parlance of their trade. The samples did, indeed, reveal high mineral content, and the following summer they returned to the hill between Emish and Anaktalak with a diamond drill rig. In October 1994, on

PROLOGUE

just their second attempt, Chislett and Verbiski drilled core yielding extraordinarily high values of base metals, especially nickel. This was Diamond Drill Hole Number Two, the Discovery Hole, and they gave the bald hill on which they were standing the grand name Discovery Hill, though the world would know the place, forever after, as Voisey's Bay. Core after core proved that Chislett and Verbiski had found a sizeable ore body of immense wealth on and beneath the surface of Discovery Hill.

In the months that followed, the Voisey's Bay discovery would be hailed as "the greatest base metals find in Canada in a generation." Soon fortunes would be won and lost over the hill between Emish and Anaktalak; the careers and reputations of powerful Akeneshau far away would be made, or broken. Besides base metals, Voisey's would also yield a wildly unpredictable tale of greed and haste, of prudence and patience, of irony and absurdity. Most of all it would produce a clash of two cultures: one ancient and patient and utterly at home in Emish and Anaktalak; the other rapacious, itinerant, and absolutely alien to the rocky, difficult coast of Northern Labrador. Both cultures, I have come to understand, are deeply imbedded in our Canadian national character.

For me, the story of Voisey's Bay began half-a-continent away on a fine Friday morning in June, in my home town of Sudbury, Ontario. At the time the events of that day didn't seem related, but I now know that the seeds of the conflict at Voisey's Bay were present, in microcosm, here in Sudbury even then, and, looking back, I regard the events of that extraordinary morning as a kind of foreshadowing...

Enter Inco

June 1995 – July 1996

1

Duelling by Degrees

FRIDAY, JUNE 2, 1995, dawned clear and warm in Sudbury—a splendid day for a university commencement. If Joan Newman Kuyek felt nervous at the prospect of what she was about to do, it certainly didn't show. Wearing a white dress beneath her academic regalia, Kuyek appeared positively radiant as she filed into Laurentian University's Fraser Auditorium surrounded by the city's social and economic elite. There was Inco chairman and CEO Mike Sopko, appearing at home and at ease amongst his equally dark-suited peers: Jim Ashcroft, Sopko's successor as Inco's Ontario Division president and a member of Laurentian's Board of Governors; Jamie Wallace, the president of Pioneer Construction, a member of the Baton Broadcasting Co. Board of Directors and chair of Laurentian's Board of Governors; and Ross Paul, the boyish, amiable president of Laurentian University. Except for Kuyek, all were blissfully unaware of the extraordinary events that were about to be set in train.

For Sopko the day must have held special promise. His wife and children were in the audience to watch him receive the university's highest honour; and he had, only days before, concluded negotiations that would see his company buy a 25 percent interest in

ENTER INCO

Voisey's Bay, the rich new nickel find in Northern Labrador. Few outside the company knew of this triumph, and Sopko had chosen that day to announce it to the world.

Even now, no one outside the Laurentian University community really knows who first came up with the bright idea to award honorary doctorates to Kuyek, one of Canada's leading social activists, and Sopko, one of the world's biggest mining moguls, on the same day. To say the two held antithetical views would be putting it mildly. And when, several weeks before the event, Kuyek expressed her unease privately to university officials about sharing the podium with Sopko, it might have given someone pause. Considering the uproar that ensued it's hardly surprising that there hasn't been a rush of volunteers eager to take the credit. But the whole thing must have seemed a Good Idea at the Time.

There were three other convocation ceremonies for the Class of '95, and the line-up could easily have been juggled. But no, Kuyek was told, she must accept her degree along with Sopko. And when Kuyek then asked that she at least receive her degree, and make her acceptance speech, *after* Sopko, alarm bells might have started ringing somewhere in the rarified circles of the university administration.

Instead, Laurentian officials simply assented to Kuyek's second request. It all sounded harmless enough. This is post-modern academe, after all, and ideology is, if not dead, then so severely discredited as to be practically toothless. So what if Kuyek got in the last word? How bad could it be?

■

Tall, trim and charismatic at 53, Joan Newman Kuyek still has about her the fresh-faced enthusiasm of the campus beauty queen, which she, in fact, once was. Winning a beauty contest during her university days in Manitoba, indeed, was a turning point in a life begun in relative affluence and privilege in one of Winnipeg's tonier neighborhoods. The way Kuyek tells it, the first original thought in her life may have occurred in the instant after her name was announced as the winner of a coveted queen title at the University of Manitoba. "I saw the heartbreak in the faces of all the other contestants, and I realized what a horrible thing competition truly is. I was supposed to be happy at the expense of these other young women and I wondered, *'What kind of system is this?'*"

After graduating from university Kuyek joined the Company of Young Canadians, a vaguely leftish, federally funded band of community organizers in its infancy, and later moved to Kingston, where she helped organize the welfare moms and ex-cons living on the wrong side of Princess Street. She was eventually elected to Kingston City Council as a progressive candidate before moving to Sudbury in the early 70's, where she has resided ever since. For over 25 years Kuyek has left an enduring legacy in Sudbury and influenced countless younger activists, helping to organize everything from the Sudbury Women's Centre to the wives of Inco workers during the titanic ten-and-a-half month strike in 1978-79. Kuyek's evident commitment and courage has won her grudging respect even among her enemies, and she have remained resolutely on the side of society's marginalized.

If Kuyek's life is Mother Teresa meets Campus Queen, Mike Sopko's is pure Revenge of the Geek meets Horatio Alger. Born into a working-class family in Montreal, Sopko has ascended the Inco hierarchy to become one of Canada's 200 most influential executives, according to the *Financial Post*, with a salary to match: $991,599 in 1994.

Just three years Kuyek's senior, Sopko was already studying for his doctorate in metallurgical engineering while the then Joan Newman was receiving her first object lesson in the true nature of competition. He joined Inco in 1964 as a junior research engineer and thrived in the competitive corporate environment, becoming president of Inco's Sudbury-based Ontario Division by the mid 80's, a senior corporate VP in 1989, director in 1991, chairman and CEO in April 1992.

ENTER INCO

If Mike Sopko, as the head of the western world's premier nickel producer, can be said to be a member of Canada's ruling class, he is definitely not *of* it. Bespectacled, portly and nerdish, Sopko does not look the part of a captain of industry, and there is still the tell-tale hint of the hard Montreal working class "g" in his speech. Although he can now afford expensive suits, his clothes still hang all wrong. A friend of mine who encountered him at Toronto's Pearson Airport remarked that the rumpled and dishevelled Sopko "looked like *a shoe salesman*, for Chrissakes."

But his appearance, like Joan Kuyek's, is highly deceiving, for Mike Sopko is a remarkable man in the annals of Canadian business. He is the first Canadian to head the Canadian-based (if not Canadian-owned) Inco, a company founded in 1902. At the time of his promotion many observers, including, it must be admitted, myself, quietly rejoiced that a native-born Canadian would finally be at the helm of one of Canada's largest and most venerable industrial enterprises. Better yet, Sopko had resided for much of his career in Sudbury, and married a native Sudburian, leading many local Inco watchers to hope that the city, and Ontario Division, would be restored to preeminence in a company whose leadership appeared distracted by ventures in Guatemala, Indonesia, and elsewhere.

While they had lived in the same relatively small city for the better part of a quarter-century, Mike Sopko and Joan Kuyek had never actually met, though they must have been aware of one another through the news media, if nothing else. These two contemporaries, who had moved so unpredictably against the grain of appearance, class, and expectation, would finally cross paths for the first time on the stage of Laurentian University's Fraser Auditorium.

■

The company Sopko came to head in 1992, it could be argued, was the first giant Canadian corporation to enter the post-industrial age, and it might be instructive to contemplate where this transformation has led.

Consider this: in the fall of 1977 Inco announced massive layoffs at its Canadian operations, even though it had made a tidy profit the year before. The country was aghast. Parliamentarians in Ottawa and Queen's Park were quick to call senior Inco executives onto the public carpet, and angry demonstrations and rallies against the company were featured as the lead item *three nights running* on the CBC national TV news. Similar announcements are routine today and rate barely a mention by Peter Mansbridge, much less a parliamentary inquiry. How, you may well ask, did we get here from there?

For much of its history Inco was one of Canada's richest, most powerful, and most hated corporations. Defiler of the air, despoiler of the environment, Inco's scattered mines and sprawling mills, smelter and refineries in Sudbury killed and maimed workers with inhuman alacrity. The company was also a direct extension of the U.S. military-industrial complex. John Foster Dulles, secretary of state under Dwight Eisenhower, and his brother Allen, the first director of the CIA, had both been Inco directors. Lacking nickel deposits of its own, the U.S. was highly dependent on Sudbury, and Canada, for its supply of what has been called "the most militarily strategic of all metals." Inco CEOs were invariably American, often former under-secretaries of the U.S. Navy, who decamped to Toronto (after Inco's corporate headquarters moved there from Wall Street in the 1970's), served out their time, and then returned to their Manhattan townhouses and summer estates in the Hamptons.

In true dialectical fashion, Inco's two-fisted, hard-nosed management style in Sudbury spawned Canada's largest and most militant union local: 598 of the Mine Mill and Smelter Worker's Union, which later became Local 6500 of the United Steelworkers of America. Inco's corporate history reached a milestone in the fall of 1978, when 10,700 Steelworkers embarked on an apparently suicidal strike against a huge nickel stockpile. Given that the strike came hard on the heels of the layoff announcement, the national media had a field day. Thousands of hardrock miners and their wives squared off against the American multinational we most loved to hate, suffering without paycheques through the longest, coldest winter in living memory, determined to bring the mighty nickel giant to its knees.

But the cruelest media cut came from a most unexpected direction: Bay Street itself. In May 1979, as the strikers staggered into their eighth month on the picket line, *Canadian Business* magazine published a cover story entitled "The Arrogance of Inco." In a monumental 20,000-word piece, reporter Val Ross traced the entire history of Inco, excoriating the company for its treatment of its unions, the environment, the nation itself.

The story shook Inco management, from the 3,000-foot level of Frood Mine to the 53rd-floor boardroom at world headquarters in First Canadian Place. The company's public image, and management morale, hit rock bottom. Harry Tompkins, Inco's public relations chief, once quipped that his job description during this period was so oxymoronic that his cronies at the Toronto Press Club nicknamed him "Titanic Tom."

■

The strike finally ended, with a union victory, in June 1979, and over the next decade Inco management introduced a raft of internal reforms that would, outwardly at least, transform the company and usher it into the post-industrial age. Language was important. "Layoffs" soon became a thing of the past, replaced by "voluntary early retirement offers." The results were nearly the same, however, as Inco led Canadian industry in corporate downsizing. The company's Sudbury workforce dwindled from 14,000 hourly rated employees in 1974 to 5,000 in 1995, while production increased by roughly half.

The Steelworkers union purged itself of militant elements and wave after wave of "early retirement offers" was warmly received by the new union leadership. The union and management decided to cooperate on safety and health matters, and the results went beyond the superficial. On-the-job fatalities dropped dramatically (though it was true that with fewer people working there were fewer opportunities to kill them), and so did serious accidents. Union members were given shares in the company, and management made a point of holding regular meetings with employees to explain the state of world nickel markets, the rationale behind certain corporate decisions, and the financial health of the company. New technology was introduced to reduce environmental pollution, and a host of visiting dignitaries, ranging from Prince Charles to Prime Minister Jean Chrétien, were trooped through company facilities to push buttons and throw switches on shiny new equipment that boosted productivity and reduced pollution, even as it eliminated jobs.

Gradually, Inco became the very model of modern corporate citizenship. Expensive ads were placed in Canada's leading slick magazines extolling Inco's newfound sense of responsibility to the environment, its workers, the community. Positive stories began to appear in the mainstream media about the way the company managed its affairs, and for a time Inco sponsored a literary section in *Saturday Night* magazine. In the 1990's Ontario Premier Bob Rae, whose party had once wanted to nationalize Inco, lauded the company as an exemplary corporate citizen. There were even reports that Walter Curlook, a senior Inco vice-president, endorsed Rae on campaign literature during the NDP's unsuccessful bid for re-election in the spring of 1995.

And yet all was not well within the far-flung corporate empire over which Mike Sopko came to preside. The difficulties began to manifest themselves in an area that had rarely been a problem for Inco—the bottom line. True, some of the causes stemmed from global factors over which management had little control: the recession of the early 90's, and the collapse of the Soviet Union that brought the free market to Russia, and Russian nickel, formerly controlled by the state, surging willy-nilly onto world markets,

ENTER INCO

further depressing nickel prices. Inco employees in Sudbury, union members and supervisors alike, grew increasingly restive under the Sopko regime. The amount of overtime demanded soared, basic maintenance was deferred in the interest of increased production, and, predictably, bottlenecks were developing, reducing both productivity and profitability.

Having led Canadian industry into the era of global competition through technology and corporate downsizing, Inco management now found itself in a new swamp. Sopko and his staff, employees were heard to say in bars, at parties, and in grocery stores, had cut too deeply. On the one hand, early retirement offers had thinned the ranks of management and senior employees who could provide invaluable knowledge and a sense of continuity. It had been many years since new, younger employees were hired. The result was an aging, exhausted workforce that was simply unable to respond to management exhortations to work harder and longer. Could it be that an ample supply of living, breathing human beings, of all ages, was still critical to profitability, even in this new epoch of high technology and the information revolution?

To make matters worse, Sopko began to publicly berate his troops in Sudbury, describing the operation there as "the highest cost producer" in the company. Most of Inco's workers, now in their late 40's or 50's, well remembered the decades when Ontario Division was Inco's lowest cost producer, bankrolling expensive offshore expansions in Indonesia and Guatemala, as well as overseeing the purchase of a U.S. battery company that resulted in one of the largest corporate write-downs in the history of Canadian finance.

Inco's internal contradictions became acute in 1994. Even as most industrial enterprises were returning to profitability, Inco's earnings lagged. While Sopko and his senior execs continued to preach cost-cutting, they themselves took home tidy bonuses. Mike Sopko's 1994 remuneration package of close to $1 million represented more than 3 percent of the company's total earnings for the year—just $29.5 million, on sales of $3.3 billion.

DUELLING BY DEGREES

In the week before the Laurentian University convocation where Mike Sopko would accept his honorary degree, I wrote a newspaper piece ruminating on these facts. Based on extensive, though off-the-record, conversations with all levels of Inco staff, the column quickly made its way onto bulletin boards throughout the company and resulted in a spate of congratulatory, if anonymous, phone calls from Inco personnel, including supervisors, who were delighted that someone had finally spoken out about what was ailing their company.

It also produced a phone call that provided the first inkling that the university's graduation ceremony would prove anything but ordinary. The caller was Inco pensioner Johnny Gagnon. An energetic 68-year-old, Gagnon had first made a name for himself as chair of the Inco Sintering Plant Committee back in the 1970's. The old sintering plant, while short-lived and only a tiny corner of the Inco empire, was a lethal hell-hole. To date, more than 180 workers are known to have died from industrial disease relating to the place, and Gagnon's campaign to have them identified and their widows remunerated has resulted in millions in compensation claims. His latest crusade was for Inco pensioners and their widows. The early retirement offers of the 1980's and 90's have resulted in two classes of Inco pensioners: older retirees, like Gagnon, receive a pension of as little as $100 or $200 a month from Inco, while more recent retirees receive as much as $2,000 or more. In 1994, Sudbury area pensioners, who may outnumber active Inco employees, opened a campaign aimed at embarrassing the company into acknowledging and redressing this glaring inequity. Now they wanted to picket Sopko's degree-receiving ceremony.

"And Mickey?" Gagnon asked, "Just how do you get to the Fraser Auditorium, anyway?"

■

The question shocked me, but it was symptomatic of the relationship between Sudbury's only university and the community it serves. The Fraser Auditorium is one of the most public spaces on the Laurentian campus, a venue for all manner of lectures, concerts and cultural events. Yet here was a working class community leader who had lived in Sudbury all his life, but had evidently never set foot in the place.

The question shocked me, but perhaps it shouldn't have. Situated on a high hill overlooking Lake Ramsey, Laurentian is tucked into the extreme southwest corner of the sprawling Sudbury Region and surrounded by affluent homes built on the shores of nearby lakes. Its white high-rise buildings give it the oft-noted, and quite literal, aspect of an Ivory Tower on the Hill. Yet Laurentian is not elitist, in the strictest academic sense. Known pejoratively by some as "Lunchbucket U" for its open admissions policy, Laurentian has educated a generation of working class kids from all over Northeastern Ontario, and the university has had a generally salutary effect on the Sudbury community during its 35-year existence.

It was not always so. If Laurentian has played an important, if curiously isolated, role in Sudbury's community life, Inco has long played an outsized role in the life of the university itself. Founded partly on Inco largesse (the company's $10 million grant to help start Laurentian in 1959 was the largest single corporate donation ever made to an institution of higher learning in Canada at that time), the university served as a cockpit for the bitter raids staged by Steelworkers on the Mine Mill union during the early 1960's. The University of Sudbury, an affiliated Jesuit college, harboured several priests who helped to organize and direct the McCarthyite witch hunts that targeted the leaders of Mine Mill, whose Communist leanings might somehow disrupt the flow of nickel into American warships and airplanes. The most prominent of the Ivory Towers is an eponymous memorial to the late Ralph D. Parker, Inco's Ontario Division president during the raids and one of the first chairs of the university's Board of Governors. Senior members of Ontario Division management have always sat on the university's governing board, and the university and Inco have long cooperated on joint engineering and mining-related projects beneficial to both.

Johnny Gagnon's modest home in the working class Valley area north of the city is a good 45-minute drive from the university in rush-hour traffic. But his question was a reminder that the psychological gulf that still separates town and gown is more than a matter of simple physical distance.

∎

Before she arrived at the Laurentian campus, Joan Kuyek, already nervous about the speech she was to give, was also concerned about crossing the pensioners' picket line. Her fears were allayed by the pensioners themselves, who assured her that they didn't intend to physically block off the campus or disrupt the ceremony. They wanted to simply mount an information picket line. They would, in any event, provide her with an escort through their line, as a gesture of respect for her years of community work and in tribute to the honour she was about to receive. They did not intend to be so gracious to Mike Sopko.

Spirits were high on the line in the brilliant sunshine, as the 40 or so picketers, in their late 60's and 70's, carried placards under the watchful eyes of two members of the Sudbury Regional Police, parked in a nearby squad car. "Twenty eight years at Inco, $108 [monthly pension]. Wouldn't it be nice if I got a little more?" read one picket sign. A TV news team had just arrived when one of the widows, a frail woman who looked to be in her 70's, collapsed onto the grass beside the university's front gate. A group of picketers rushed to her assistance and helped her back onto her feet before she collapsed a second time. One of the police constables, who also rushed to her aid, had seen enough. She quickly grabbed for her radio and called for an ambulance.

Probably, I told myself, the woman had been overcome by the hot sun. Still, Gagnon and others had insisted that some Inco widows had been reduced to eating dog food to stave off malnutrition. The picketer, whose prostrate form was still on the ground as I walked up to the Fraser Building, appeared to be extremely thin and weak. Could it be she was actually malnourished?

■

In any event, the picketers never did confront Mike Sopko, who was whisked onto the campus in a vehicle with tinted windows driven by a member of Inco's own security department. Other members of the department were visible inside the Fraser, wearing sports jackets and ties with earphones plugged into their ears, Secret Service-style. The 800-seat auditorium was filled to overflowing, with graduates sitting nearest the stage, their families and friends seated behind.

As planned, Sopko spoke first. To an ear admittedly untutored in the genre of convocation addresses, his speech was at once sincere and pedestrian. He confirmed publicly, for the first time, that Inco was about to invest in the elephantine base metal discovery at Voisey's Bay, an announcement that was fleshed out five days later on Bay Street. He congratulated Sudbury for its economic diversification away from mining, a remark inadvertently relevant to the furor that was about to follow. In a "to you the torch is passed" sort of reference, he warned the graduates of the rapidity of change in the modern world. Knowledge, he intoned, is increasing exponentially, and the human species will soon have learned as much in one year as it had during the first 20,000 or so years of evolution. ("Did he mean knowledge or *information*?" I overheard one Laurentian academic muse aloud at the post-ceremony reception.) His duty done, Sopko received polite applause, was presented with his degree, and sat down.

At about this point in the proceedings the fax machine began humming in the nearby office of Janet Sailian, the university's director of marketing and communications. With an eye evidently cast toward her employer's largest corporate donor, Sailian, on behalf of the university, was issuing a disavowal of the pensioners' picket line to every newsroom in the city, "regretting that this celebration has been disrupted by a protest unrelated to the university and which detracts from the focus on our graduands." The septuagenarians on the picket line might have been forgiven for wondering how it was that they, who had survived the sintering plants, the Ralph D. Parkers, and the raids, who had built the union and the company and the city, whose taxes had supported the university, whose labour had provided Laurentian with its initial $10 million endowment in the first place, could ever be unwelcome at the university's gates, to express whatever opinion they might have, whenever they wanted. It was an egregious gesture that did little to close the metaphysical gap between John Gagnon's bungalow in the Valley and the Ivory Tower on the Hill.

Back in the Fraser Auditorium, Joan Kuyek stepped up to the dais, and took a deep breath. "Every morning early," she began, "I go running with our dog Blue in the woods on the corner of Lasalle Boulevard and Frood Road . . ." What followed was a synthesis of Marxism, feminism and Aboriginal spiritual teachings, and ten of the most publicly humiliating minutes that Mike Sopko, Inco, and patriarchy would ever suffer. Inco's security forces might protect their boss from unwanted confrontations with pensioners, or sudden physical altercations with disgruntled employees at the Fraser Auditorium, but nothing could shield him from the indictment that was to follow.

For me, one of the most moving passages of Kuyek's speech to the young graduates was the observation that training was one thing, but that what one does to earn a living with that training is something else. "Twenty years ago I worked for Bell Telephone as a service representative, where part of my job was disconnecting telephones for non-payment of the bill. One day, I was faced with disconnecting a woman with no car, whose husband was unemployed, whose child was sick and who lived out of town. They only owed three months regular service—no long distance. She pleaded that if the phone was disconnected her husband would never be able to find work and she would be unable to get help for the child. The rules said I was to disconnect her. I broke down in the middle of the call, but my supervisor took over and did it without a qualm. She was 'well trained.'"

Perhaps I was projecting, but in the slight pause that followed that line I could almost feel the moral point screwing itself home to the graduating class, many of whom were young women.

As Kuyek continued her address, Mike Sopko, who was facing the audience from the front row on the stage, began to colour visibly. His face involuntarily assumed a sickly grin, and his head wagged in denial from side-to-side. At one point Sopko turned around and rolled his eyes at Ontario Division president Jim Ashcroft, seated directly behind him, and Ashcroft leaned forward to commiserate with his boss. Finally, Sopko began to whisper earnestly and at some length in the ear of university board chair Jamie Wallace, who listened attentively, ignoring Kuyek's speech in full view of the assembled multitude.

A moment of stunned silence followed the conclusion of Kuyek's remarks. One observer reported seeing Laurentian University president Ross Paul flash Kuyek a thumbs up sign as he moved to escort her away from the podium, a spontaneous gesture that, if made at all, Paul might well have regretted later. And then, in fits and starts, it began, the greatest honour of all, and there was nothing equivocal about it. The vast majority of the audience was *standing*, applauding Joan Kuyek, while Sopko and Ashcroft and Wallace sat, ashen and immobile.

Canada's captains of industry, like those everywhere, have private clubs and golf courses, security guards and chauffeur-driven limousines, to protect them from the remonstrance of the envious, the resentments of an ungrateful public who might rarely but truly grasp that it has taken the suffering of the many to produce the fortune, and the privileges, enjoyed by the few. But in that one rare moment in the Fraser Auditorium, when almost everyone was clapping and standing, it was clear that Joan Kuyek had at last hoisted Inco on its own historic petard.

∎

Events in the aftermath of Kuyek's speech moved swiftly, often furtively, and soon assumed the status, in Sudbury at least, of urban myth. At the reception that followed the ceremony Mike Sopko studiously avoided Joan Kuyek. The fellow honorary doctorates never shook hands, and no joint pictures were ever taken. When a bemused Kuyek mentioned this during a Kodak moment outside the reception, her son Devlin, a handsome, strapping undergraduate at Queen's, looked down at his mother, his eyes beaming with merriment and pride: "Well mom, what did you *expect* the president of Inco to say?"

On the night of the ceremony, the university's elite gathered for a reception at the home of the university president, an improbable faux Spanish pile, complete with stuccoed facade and Moorish trim, overlooking Lake Ramsey. Members of the university's Board of Governors were there, as well as the recipients of honorary degrees and senior university administrators. Host Ross Paul and one or two other dean-level figures were gracious to Joan Kuyek, but the others shunned her like the plague, making her feel, as Kuyek would observe the next day, "about as welcome as a skunk at a garden party." At one point she fled the reception so that no one would see her tears.

On Monday morning a peculiar witch-hunt began at Laurentian University. Certain professors who had been on the stage were accused of having started the standing ovation, of causing the "impressionable" graduates to follow suit. If a videotape of the event had existed, it would probably have been played over and over in slow-mo to determine *"who* stood *first?"* Most of those called on the carpet, being mature and erudite individuals, laughed the whole thing off. After all, wasn't that why a university existed? For an exchange of intellectual views?

On Tuesday morning, an emergency meeting of the university board of Governors was convened. On the agenda was the Kuyek speech, and how the university should respond. Rumours circulated that Ross Paul's job was on the line because of his alleged thumbs up affront to Inco (Paul later emphatically denied ever making such a gesture); that Inco and some faculty members and governors were demanding an apology; and that Inco was canceling a multi-million-dollar contribution to the university's fund-raising campaign. The closed-door meeting, which lasted well into the afternoon, ended with no public explanation for its having been called. Ross Paul remained the president of Laurentian University, and no public apology for the commencement was ever tendered. Paul's executive assistant, who made the arrangements for Joan Kuyek's presentation, resigned over the summer.

My suspicion is that most members of the university's pluralistic board would have opposed any form of recrimination, especially a public apology, on the grounds that it would only deepen and prolong the university's embarrassment. Some faculty members, I'm told, threatened to resign if any apology was made.

It is significant, however, that within a month of the graduation ceremony, Ross Paul, in his regular column in the *Sudbury Star*, cited Mike Sopko, Jim Ashcroft and Jamie Wallace as the kind of public-spirited, forward-looking businessmen who had made Laurentian such a great university.

Both Laurentian and Inco issued a public statement, after repeated media inquiries, to the effect that Inco's support of Laurentian would continue as before. Neither would offer specific dollar figures, causing a number of Sudbury editors to wonder just how much money Inco gives to the university each year, anyway. Neither the company nor the university would answer that question, either.

Although I faxed the text of Kuyek's speech to CBC's *As It Happens* offices in Toronto a few days after it was delivered, it was two weeks before a producer phoned back, apologizing for the delay in her response. *Two weeks?* In this age of the Information Superhighway? The Pony Express would have been faster. It would take nearly six months for this story to reach beyond the bounds of the Regional Municipality of Sudbury. Individual acts of defiance and courage like Joan Kuyek's are doubtless happening every day in this country, but because they run counter to the corporate agenda few of us will ever know about them.

■

Joan's speech had referred to "the values I have learned to name from my Anishnabi elders." I sat beside two of those elders, Art and Eva Solomon of Sudbury, during the commencement. A highly respected Ojibway elder, the holder of an honorary doctorate from Laurentian, Art's pride in Joan's speech was palpable even though he was, at 80, using a wheelchair and in extreme discomfort from a kidney ailment.

Thinking they would want to know the latest, I phoned Eva Tuesday afternoon with news of the board of Governors meeting. When I mentioned the rumoured possibility of a public apology, there was a momentary pause at the other end of the phone.

"Oh," Eva said softly in a slyly plaintive tone of voice. "Poor Inco."

2

Rosencrantz and Guildenstern 'r Us

ON THE DAY AFTER the Laurentian board's *in camera* meeting, Mike Sopko announced in Toronto that Inco was buying a 25 percent interest in the Voisey's Bay property at a cost of $750 million. Inco was becoming a partner of Diamond Fields Resources Inc., the company that owned the rights to Voisey's Bay, and of its owner, the brash stock promoter Robert Friedland. Bay Street, generally, applauded. But at a mining conference in Sudbury in November 1995, lawyer Steve O'Neill approached Sopko privately with a warning about Inco's new investment in Northern Labrador.

A Sudbury-based expert on Aboriginal land claims and treaty rights, O'Neill urged Sopko to remember that the minerals Inco had spent so much on were, in fact, on someone else's land—namely, the Innu and Inuit peoples of Northern Labrador. Sopko's response, O'Neill would say later, was strictly non-committal. But by July of 1996, the Inco CEO's reaction became clear enough. He liked Voisey's Bay so much HE BOUGHT THE WHOLE DARNED COMPANY! But not before Inco's cross-town competitor, Falconbridge, stunned the Canadian financial world by coming within a whisker of stealing the rich nickel deposit right out from under Inco's nose.

■

The whole affair made Sudburians feel a little like Rosencrantz and Guildenstern, those hapless characters from *Hamlet* who moved on the margins of history, and of the play, forever at the mercy of much larger forces they could barely glimpse, much less understand. So it was in Sudbury, where we suddenly found our fortunes and futures held hostage to a place called Voisey's Bay, Labrador. We were forced to sit on the sidelines and watch while corporate titans wielding billions of dollars played a high-stakes poker game with money originally earned from the ground beneath our feet. And, make no mistake, we were, like Rosencrantz and Guildenstern, absolutely, utterly helpless.

The financial press was too polite to say it, but Inco got royally snookered by Robert Friedland and Falconbridge in the play for control of Diamond Fields Resources Inc., the company that still owned 75 percent of Voisey's Bay.

The 45-year-old Friedland was Inco's erstwhile partner in Voisey's Bay. Until late January 1996 that is, when Friedland said he'd accept an offer from Inco's crosstown rival. Falco was willing to pay $4 billion for his remaining interest in Diamond Fields. Friedland would pocket a cool $500 million in cash for himself.

What a guy, this Friedland. He "got early business experience in the field of LSD trafficking of which he was convicted at 19," the *Globe & Mail's Report on Business* noted. Since then he has shunned any responsibility for "a leaking swamp of cyanide, acid and metallic wastes" at a mine he once owned in Colorado. U.S. taxpayers were stuck cleaning up that mess at a cost of $110 million.

He struck me as just the kind of character you'd want to invite home to dinner, maybe introduce your sister to. Or, better yet, go into business with. Maybe become partners! And that's just what

Inco had done. Everything seemed copacetic until the afternoon of Friday, January 26th, 1996 when Inco presumably learned, along with everyone else, that they'd just been sold down the river. Friedland, it appeared, was about to sell the 75 percent of Diamond Fields that Inco didn't own to its bitter rival Falconbridge.

How did Bay Street react to Friedland's double-cross? They loved it! "He's a genius, isn't he?" marveled Norman Keevil, CEO of Teck Corp. "Robert Friedland is a warrior, an entrepreneur warrior," gushed Walter Berukoff, president of Miramar Mining Corp. "He's done a prince of a job," agreed Toronto mining promoter Patrick Sheridan.

It would have been tempting to laugh at this madness, were it not for the potentially serious, indeed downright dire, consequences that all this appeared to hold for the non-genius, non-warrior, non-prince Rosencrantz-and-Guildensterns in Sudbury. The Falconbridge bid for Voisey's forced everyone to take a hard look at what the new ore body could mean for Inco, which, as part of its 25 percent acquisition, had the exclusive rights to sell Voisey's Bay ore for the first five years, probably beginning in 1999.

"They absolutely have to market the metal, period, full stop," Friedland told the *Globe*, "Voisey's Bay will rule."

"In several years [Inco] may have to cut its own production and lay off workers in order to live up to its obligations under its marketing agreement with Diamond Fields," reporters speculated. Gulp.

Under the robust conditions of early 1996, world nickel markets could probably have absorbed the Voisey's Bay output with little impact on production in Sudbury. But when the cycle turned, as it always does, Inco would still have to sell Labrador ore, which would be much cheaper to produce than the Sudbury product.

Sudbury residents were left wondering where cutbacks were most likely to occur in the event of world oversupply. The deal did not spell the imminent demise of Sudbury's nickel industry any time soon. So long as millions were still being spent in Northern

Ontario by Inco and Falconbridge for exploration and development and by Falconbridge for its smelter retrofit to process ore from the Raglan deposit in Northern Québec, it was a safe bet the nickel giants would still be in for the long haul.

But Sudburians were urged to keep a close eye on big-dollar developments like the Victor mine and McCreedy East at Inco and the smelter and Edison Building expansion at Falco. If these were scaled back it could be an ominous augury for long-term intentions.

To sweeten its offer for Voisey's, Falconbridge reportedly promised the Newfoundland government that it would mill, smelt **and refine** all ores within the province. Refining, of course, is something that Falconbridge has never done in Canada during its 60-plus years of mining here. All smelter concentrate (known as "matte") has been shipped to Norway for refining and eventual sale into the European market. Inco, to its credit, has refined ores in Ontario since opening the Port Colborne refinery in 1917. By allowing Falconbridge to ship ore in its unrefined state, successive Ontario governments of all three political parties have allowed jobs and potential added value to be exported. That, apparently, wouldn't be happening at Voisey's Bay, and I was glad. I was happy, too, for our brothers and sisters on the Rock. If ever there was a province in dire need of the high-paying industrial jobs like those the Voisey's deposit would produce, it was surely Newfoundland.

Although the new Voisey's Bay discovery seemed certain to further erode Sudbury's importance to world nickel markets, certain opportunities loomed, too. We were still a global centre of expertise in base metal mining, smelting and refining, and it seemed reasonable to expect that a fair number of Sudburians would be moving down east to help get the project up and running, which might, in turn, create some job vacancies back in Sudbury. There might also be sales opportunities for Sudbury's burgeoning mining-equipment manufacturing sector.

Finally, it seemed there might well be a joker in the Voisey's Bay deck that Inco and Falco had never faced in its other Canadian operations—Aboriginal land claims. The Voisey's ore body was on land claimed by both the Innu and Inuit peoples, and such matters are ignored at one's peril, as the Québec government learned when it tried to bring another James Bay hydro development on stream.

Perhaps the Innu and Inuit would be mistrustful of the social and environmental consequences of the Voisey's Bay project and wary of entering into agreements to facilitate development. Or maybe they would just think a white businessman's word can't be trusted. And, given Robert Friedland's track record, and the business community's adulation of it, who could blame them?

■

In the end, Inco won the bidding war over Voisey's Bay. The deal gave Inco full ownership of Voisey's for $4.3 billion Canadian.

That summer, corporate projections for Voisey's were positively ebullient. Production would begin as early as 1998. The ore was so rich that the cobalt and gold would pay for the nickel, making it virtually "free." And spirits in Sudbury sagged. Worries about the impact of the Voisey's development on the town's long-term future began to dampen everything from real estate values to advertising in the local media.

Nor was the anxiety groundless. Sudbury is, after all, still a mining town, and everyone knows the fine line that separates rock, which is worthless, from ore, which is valuable. Everyone in a mining town, too, knows the transitory nature of the industry. From the first ounce of ore extracted the town's economic raison d'être begins to dwindle. All ore bodies are finite, and subject to competition from other, newly discovered ore bodies.

This was a lesson Sudburians knew all too well. They had watched as Elliot Lake, once the "Uranium Capital of the World," became completely superannuated, courtesy of richer uranium deposits in Northern Saskatchewan. One by one Elliot Lake's uranium mines had shut down permanently during the 1990's. Elliot

Lake is only a hundred miles from Sudbury, and many mining families have moved back and forth from one camp to the other. Would Sudbury go the way of Elliot Lake, courtesy of the fabulous new discovery at Voisey's Bay?

No city in Canada had more to lose than Sudbury in the Voisey's Bay play. And yet, while the national news media were filled with stories about the positive impact the development would have on Newfoundland and Labrador, scant attention was paid to the negative impact it could have on Sudbury, and Ontario.

Against this anxious local background, I persuaded *Northern Life* publisher John Thompson and Managing Editor Carol Mulligan to send me to Voisey's Bay. No Sudbury reporter had gone there yet, and who knew? Maybe I'd turn up something that might help reassure Sudbury's business community. And so it was that on the morning of Tuesday, October 22nd, 1996, I found myself on the approach to the dirt airstrip of Nain, Labrador.

A Trip to Voisey's Bay

October – November 1996

3

Welcome to Nain

THERE'S AN UPSIDE and a downside to occupying the seat right behind the cockpit of the plane that takes you into Nain, Labrador. The upside is that the first row of seats inside the cramped and spartan cabin of the Twin Otter offers more leg room. The downside is that you see, through the cockpit window, what the pilots see on their approach to Nain—a dirt airstrip that looks about as long as a fairway at the local Mini-Putt (on a par two hole).

The Nain airstrip is the stuff of Labrador legend—at just 600 metres the shortest strip on the Labrador coast, plagued by downdrafts, reached by threading, not over, but through the rocky peaks that surround the town, and banking steeply out over Nain Bay before making a final approach. Since the discovery of the Voisey's Bay nickel deposit and the claim-staking rush that followed, the airstrip is reputed to be the busiest in Atlantic Canada, or Labrador, or the Labrador coast—take your pick. Where once there were one or two scheduled Twin Otter flights a day into this formerly sleepy Inuit settlement of 1200, there are now (weather permitting) five or six.

That doesn't count the incessant chopper traffic, ferrying workers and supplies to the Inco property at Voisey's or carrying diamond

A TRIP TO VOISEY'S BAY

drill crews for a dozen junior mining and exploration companies into the vast hinterland that surrounds Nain, all of which is considered to be part of the rich Voisey's play.

You choose not to watch the final, dizzying descent onto the airstrip, and to ignore the faintly crabwise motion of the plane as the wheels touch down on the dirt, flaps full down, the props roaring, in an effort to stop the Twin Otter short of the fast-approaching waters of Nain Bay at the end of the strip.

"Gee, I feel like giving these guys a hand," I marvelled to my seatmate as the plane finally rolled to a stop.

"Naah," replied Gail Stanley, the head nurse for all of Labrador, and a veteran of northern aviation, with a dismissive wave of her hand.

"Just another day at the office for them, I guess?"

"Just another day at the office," she affirmed.

Welcome to Nain, Gateway to Voisey's Bay.

■

Although the trip from Sudbury had taken five flights over a full day-and-a-half to reach Nain, with an overnight stop at Goose Bay, I still had not made it to Voisey's Bay—not quite. There is still one more hop, a 30 kilometre flight to the south by helicopter into the mine site proper. Stewart Gendron, formerly of Inco's Ontario Division in Sudbury and now president of Voisey's Bay Nickel Company Ltd. (VBNC), Inco's Newfoundland subsidiary, had already told me that there would be no way I could visit the mine site.

"Mick, there's just too much going on right now," he had said regretfully from his St. John's office by phone on the eve of my departure from Sudbury. So I resolved to spend my first 24 hours in Nain seeing what I could learn about Inco's operations from a distance, and staring wistfully south, over the cold blue-gray waters of Nain Bay.

What I learned was this: Voisey's Bay is, first of all, something of a misnomer. The mineralized zone itself is in fact several kilometres inland, centred on what is now known as "Discovery Hill." The Discovery Hole, which turned Newfoundland prospectors Chris Verbiski and Al Chislett into millionaires overnight, was drilled on October 22, 1994, two years to the day before my own arrival in Nain.

(More local legend: it is said that Verbiski and Chislett, unable to contain their excitement, confided the significance of their discovery to the chopper pilot who flew them out that October. He promptly invested heavily in their company, Archean Resources, and became a millionaire, too.)

The Discovery Hole, atop a hill overlooking Voisey's Bay, proved to be the centre of a massive sulphide mineral deposit right on the surface, which has now become known as The Ovoid. Subsequent drilling revealed that the Ovoid, projected to become the site of Inco's first mine, an open pit, contains 31.7 million tonnes of ore, grading 2.83% nickel, 1.68% copper and 0.12% cobalt.

These assay results, while quite rich, are in fact comparable to Sudbury ores—in their infancy. *The Report of the Royal Ontario Nickel Commission of 1917* found hand-picked ore at Creighton Mine grading 4.44% nickel and 1.56% copper. Today, Inco still has many million of tonnes of ore left in its Sudbury reserves, but most of it is at far greater depth than the Ovoid deposit, and therefore far more expensive to extract. Further drilling revealed a second huge ore body, the Eastern Deeps, which, while larger, is at greater depth and getting at it will require underground mining methods for extraction. The Eastern Deeps is currently estimated to contain an additional 50 million tonnes of ore, grading somewhat lower than the Ovoid—1.36% nickel, 0.67% copper and 0.09% cobalt—grades not greatly superior to Sudbury's remaining reserves.

Yet another zone of high-grade mineralization has been discovered to the west of the Discovery Hole. Dubbed the Western Extension, its composition was still being explored at the time of my visit. Suffice it to say that Inco has ore reserves in the Voisey's Bay deposits that will last for decades to come.

■

While in Nain I was also able to gain an overview of the geography and topography of Voisey's and its surrounds. The area reminded me of Coniston, a town just to the east of Sudbury that had been devastated by smelter fumes between the 1920's and 1972. There are the same rocky, treeless black hilltops, only far higher and larger, rising 1000 to 3000 feet above the tidewater of the Labrador Sea.

There are trees—scrubby black spruce and tamarack—in the lower elevations, and in the town of Nain itself. The hills rise abruptly from the water's edge of the ragged Labrador coastline, indented with innumerable bays, islands, and fjords.

The Voisey's Bay area, it turns out, is actually a peninsula jutting eastward into the Labrador Sea at roughly 56 degrees 22 minutes north latitude. It is on a line far to the north of Edmonton, but south of the 60th parallel, which divides the Northwest and Yukon Territories from Canada's Western provinces.

Inco, I was told, had two camps in the area, one at Voisey's Bay, the second at Anaktalak Bay, on the other side of the peninsula. In all, there were perhaps 300 personnel on site, with Anaktalak being the larger of the two camps. The company's plans called for an open pit mine to be established at the top of Discovery Hill, and a 20,000 tonne per day mill to be located nearby. Ore concentrate will be moved north, eight kilometres downhill to Anaktalak Bay, where a deepwater loading facility and permanent camp will be built. Tailings from the mill will be pumped to the north and west of the Ovoid, into a small landlocked basin. In this way, it is hoped, leaching into the local watershed and thus into the Labrador Sea will be prevented.

Inco also plans to build its own airstrip on the top of Discovery Hill. At 1250 metres, it will be almost twice as long as Nain's, a plan of which I heartily approve. Because the local Aboriginal peoples objected to roads being built without a full environmental assessment, I was told that the two camps, Discovery Hill, and the myriad of diamond drill rigs that dotted the peninsula were accessible only by helicopter or by water. Inco was reported to have leased a fleet of ten or eleven machines from Canadian Helicopters Eastern out of Goose Bay for just that purpose. Big bucks. What's more, the company had also leased a Canadian Coast Guard icebreaker, the *Sir John Franklin*, and anchored her for the season just offshore in Anaktalak Bay, to serve as a floating hotel for the overflow of workers who could not be accommodated at the campsite.

(I found the ship's name somewhat ominous: why would anyone name an icebreaker for the leader of the most disastrous expedition in the history of the Canadian Arctic? But I digress.)

I was, at the end of my first day in Nain, more eager than ever to see these sights, and the millions upon millions of dollars they represented. I was abashed, that first night, to encounter a competitor, Graeme Hamilton, the Atlantic correspondent for Southam News. He informed me that he was due to fly into Voisey's for a tour aboard a company helicopter the next day. Inco wasn't too busy to squire around a reporter for a news chain recently purchased by Conrad Black, but no such possibility existed for an independent reporter from Sudbury. Was it something I wrote? I was further abashed to learn that Inco based its own choppers in the camps, and not at the Nain air strip, thus scuttling my hope that I might hitch a ride with an unsuspecting pilot. I fell asleep that night with difficulty, excited to have come so far and to be so close to my goal of reaching the vaunted Voisey's Bay nickel deposit, but worried that the last 30 kilometres might prove the greatest distance of all.

4

Voisey's Bay at Last

"Oooooh, it smells just like the bathroom after Gerald's been in it," giggled Cynthia. And indeed the boat *was* filled with the unmistakable odour of feces from the dying seal that Richard Pamak had just shot.

"Boy, you're in shit with Brigitte Bardot now," I teased Richard as he lashed the 80-pound seal to a running light stanchion.

"To us it's a way of life," Richard bristled. At once I regretted the remark, and told him so, because I knew what he said was true. His prey secure, Richard returned to the wheel of his boat, and the twin 70 horsepower motors roared to life.

I had met Richard Pamak only hours before at the Nain airstrip, inside the tiny hut beside the airstrip, where he works as the local weather man. I had hired him, and his boat, to take me to the main Inco campsite at Anaktalak Bay, some 40 minutes to the south of Nain by speedboat.

I had liked Richard at once. Although only in his early 30's, Richard is a former mayor of Nain and typical of the younger generation of Inuit who have gone "out" for advanced schooling—in his case to Memorial University in St. John's.

With us on our foray down the coast was Jacko Merkuratsuk, a friend of Richard's, and three young offenders (we'll call them Gerald, Cynthia and Jonathan) from the group home where Richard worked after his day job at the weather office.

I'd been forced to travel to Voisey's by boat not only because Inco had told me I wasn't welcome at their site, or aboard one of their leased helicopters, but because a heavy fog had grounded all air traffic in and out of Nain for the day, anyway. But the same windless conditions that left the fog hanging immobile several hundred feet overhead also gave rise to ideal boating conditions along the coast.

I had already fallen in love with the stark beauty of the North Labrador coast, and was eager to see more of it. But I was also, I knew, seeing it at its most benign. Just the week before an early winter storm had dumped 18 inches of snow on the area and lashed Nain with winds that had driven a number of boats ashore on the Bay, destroying several of them.

Wind is a constant weather factor in Northern Labrador. The previous January and February, Richard had explained inside his snug office, the average daily low temperatures, with the wind chill, had been -39, with a maximum low of -66. Air temperatures had averaged -28 to -30 degrees Celsius during the same period. I thought about trying to operate an open pit mine in such conditions, and wondered if miners and management alike wouldn't yearn for the warm bowels of Sudbury's Creighton Mine, after all.

■

"We'll shoot a few seals on our way, if it's okay with you," Richard said before we had even cleared Nain Bay. He had emerged from the cabin with a .22 rifle equipped with a telescopic sight.

"Go for it," I replied. And so I found myself in the middle of a seal hunt.

After the first kill we were soon whipping across the smooth waters of the Labrador Sea at 45 mph. The clouds lowered over us at

200 feet or so, obscuring the rocky peaks that surrounded the bays, islands and inlets that stretched out before us.

The first point of local interest on our itinerary was Ten Mile Bay, where the Inuit, under the auspices of their own Labrador Inuit Development Corporation, operate a labradorite, or soapstone, mine. Located just outside of Nain, and staffed entirely by Aboriginal management and labour, the Inuit were blasting away a mountainside and shipping its contents to Italy. I couldn't help noting that the quarry is located right on the water, in contrast to the open pit at Voisey's, which will be inland, comparatively speaking. But then the Inuit operation also doesn't present a pollution hazard in the form of mine tailings.

As Ten Mile fell behind us Richard described the type of seal he'd just shot, and explained there are five different species to be found in the waters around Nain. Even from a great distance, he could instantly identify the species. It was easy, he explained, just look at their behavior in the water. "When I was a kid, growing up in Hopedale (an Inuit community to the south), my dad would take me out and we'd shoot maybe two seals in an hour. The skins were worth $75 apiece, not bad money for a few hours' work. Now, that skin isn't worth a penny." Thanks, I knew, to the activities of the anti-fur lobby in Europe. Despite a small rally in fur prices recently, the trapping industry has pretty much collapsed as a source for Inuit cash revenue, as has the fishery. The cod stocks began disappearing from the waters of Labrador twelve years ago, I was told by a Fisheries officer. Country food, however, remains abundant—seal, caribou and partridge. Caribou are so plentiful that in recent years they have even been seen in Nain and once in a great while have to be cleared from the airstrip before the arrival of incoming flights.

Occasionally we'd pass by a flatter piece of coastline with a rocky beach, and Richard would point out a fishing/hunting camp—little more than a small shack, usually, but shelter from the elements nonetheless. At last we rounded an island and Richard shouted over the noise of the engines. "There's the *Franklin*." Sure enough, I could

A TRIP TO VOISEY'S BAY

see a small red shape far ahead—the *Sir John Franklin*. Richard piloted skillfully around the giant vessel, which towered over us, while I snapped pictures. A few curious onlookers on the deck of the *Franklin* waved at us, and we waved back. Several minutes later I jumped over the side of Richard's boat and onto the shore of Anaktalak Bay. "Richard, I love ya for this, man."

"We're going to hunt a few more seals, we'll pick you up in half-an-hour, all right?" I could hear the pop of the .22 before they were even out of sight.

It was now nearly five o'clock. I had arrived in the middle of rush hour, Anaktalak-style. Perhaps 20 men, and a few women, stood on the beach looking out at the *Franklin*. They were waiting for the chopper I'd seen on the stern of the ship to come and take them "home" for dinner.

"Why not use a motor launch?" I wondered aloud. They were only a few hundred yards from the big ship, after all. But no one had a coherent answer. This is the way things are done when cost is no apparent object. Even loaves of bread, I was told, arrive at Inco's camps by chopper. I strode up the beach toward the camp, a maze of interconnected ATCO trailers. To my left I saw half-a-dozen more helicopters, and crates stacked high containing snowmobiles, apparently awaiting the onset of winter.

After quickly snapping some pictures I headed for the Main Office, where, a sign informed me, all personnel were to check in. I was greeted by a mildly curious but congenial security guard. I explained who I was and where I was from. Did anyone from the company know I was there? Oh yes, I assured him, Voisey's Bay Nickel's newly hired Aboriginal Affairs Advisor Mike O'Rourke, who was still in Nain awaiting good flying weather, knew I'd come. Which was true. He moved to phone O'Rourke for confirmation. "It's okay," I said hastily, "I've gotta meet my ride back, so I can't stay long. But I sure could use a washroom."

"Just around the corner," he motioned. "Would you like to sign our guest book?"

"You bet," I replied, and triumphantly wrote , "MICK LOWE, NORTHERN LIFE, SUDBURY, ONT."

The camp, I quickly concluded, was like any other semi-permanent northern outpost, equipped with electricity, running water, clotheswashers and dryers, pool tables, and a satellite dish. It had all the comforts of home except, well, home. While the trailers doubtless encompassed an impressive amount of square footage, I found the whole place claustrophobic. I do not envy the mine and mill workers who will find themselves working 12 hour shifts here, 14 days on, and 14 days off, although I suppose more spacious quarters will eventually be built.

■

My ablutions completed, I was just as happy to be back on the beach, scanning the water anxiously for Richard's return. It was starting to grow dark, and lights were winking on board the *Franklin*. Chopper after chopper roared into the twin pads on the beach, unloading diamond drillers and government bureaucrats and company officials from their day's work in the bush, picking up still more workers for the short hop to the icebreaker. They hurried to and from the Jet Rangers, bending low beneath the prop wash, the takeoffs and landings marked with terrific blasts of air that blew bits of sand and gravel into my face.

The light was definitely beginning to fade when Richard hove into view at last. He was alone. "I thought maybe you'd forgotten about me."

"We just went around the corner to clean the seals. We got another one."

We hastily motored around a point to collect Jacko, Gerald, Cynthia and Jonathan. They were on a low rocky outcrop, the seals cut up and placed in three piles—offal in one, skins in another, meat in a third.

"Next time, boys, don't cut them up," Richard said with a grin, explaining that Gerald and Jonathan had never cleaned seals before. The meat was quickly stuffed inside two plastic garbage bags and hefted aboard.

"Now Gerald will drive us home, eh?" Richard grinned again. Gerald shrugged and smiled in embarrassment.

"You have to know where you're going by looking back to where you've been," Richard said evenly. "Out here, you don't just go for a boat ride."

Jacko took the wheel, trimmed the engines, and soon a white wake was fanning in perfect symmetry from beneath our stern. Richard and Jacko peered intently into the gathering dusk as we roared back to Nain, while the three kids shivered in the open cockpit, trying their best not to show how cold they were. At one point, Cynthia laughed uproariously at something either Jonathan or Gerald had said. I gathered I was the butt of the joke, but never did determine what it was. My karma, I guess, for the crack about Brigitte Bardot. No matter. I was positively exultant in the freedom afforded by these waters, by the harsh beauty of this land. And the image that will remain for me of this trip, long after the memories of the helicopters and diamond drills and the reach of almighty Inco plopped discordantly into the awesome, intimidating bush of Northern Labrador have faded, will be this:

Of Richard and Jacko on that night run back to Nain, their faces above the windshield in the perishing wind, apparently oblivious to the cold, their hair streaming back, proud, vigilant and utterly at home in this awesome, intimidating land. Proud and vigilant and free.

5

Who Owns Voisey's Bay?

THE MEETING taking place before me in the hotel dining room in Nain is the talk of the town, and illustrative of the complex political environment Inco has entered through its purchase of Voisey's Bay. Perhaps two dozen Inuit elders and their advisers are just sitting down to dinner. The food is incidental. On the real menu: a high-stakes diplomatic meeting of two nations bent on determining who owns the rights to the vast mineral treasure of Voisey's Bay.

"Wait a minute," I hear you say. "Didn't Inco pay more than $4 billion for Voisey's Bay? Surely Inco owns it, fair and square." In the white man's world you'd be right. But this is Nain, the northernmost settlement of Northern Labrador in the Canadian sub-Arctic, and it's a different world altogether. It's also just 30 kilometres north of Voisey's Bay. After a brief prayer the elders are seated at half-a-dozen tables. The waitresses hover around the diners, nervous, respectful, and worried, as one of them confided later, that she might commit a *faux pas* that would spark a major diplomatic row. There is a sense of great dignity and gravity to the gathering before me, and it really is a meeting of two nations. The hosts are the elders of the Labrador Inuit Association or LIA, the organization representing 4500 Inuit people scattered through five communities of coastal Labrador, including Nain. The visitors, who have just arrived on a special charter flight, are Nunavik Inuit elders from the Québec side of the Québec-Labrador peninsula. Their business wing is called the Makivik Corporation, and they are led by headman Zebedee Nungak.

The Nunavik elders have flown here to Nain in an attempt to explain to their Labrador kin why they are about to lay claim to a vast area of northeastern Labrador already claimed by the LIA, an area that just happens to include Voisey's Bay and the minerals beneath it. In the lingo of the land claim business this is known as "an overlapping claim," and it's not the first. Besides the LIA, and now Makivik, the Innu Nation also claims Voisey's Bay. Two things are certain in all of this: neither the Inuit nor the Innu has ever signed a treaty or in any way extinguished title to this huge "empty" land that has been occupied by their ancestors for centuries, if not millennia. And it does seem clear that both the LIA's people and the Innu have hunted, fished, trapped and camped right on Voisey's Bay for as long as anyone can remember. Call me simple-minded, but if I'm Inco I'm just a tad nervous at having paid $4 billion-plus for something that might not have belonged to the seller (Robert Friedland and Diamond Fields Resources) in the first place. No one from the company or the Newfoundland government will admit it, but the land claims question might just put a stick through the spokes of the rich Voisey's Bay development. But these elders, making speeches now in Inuktitut, they know it. The land itself is Aboriginal, not just the people who live there. Hundreds of millions, perhaps billions of dollars, are hanging in the balance of unresolved land claims at a series of dinners and meetings just like this one.

■

WHO OWNS VOISEY'S BAY?

The Nunavik Inuit and two lesser contenders aside, there are really only two serious claimants to Voisey's Bay: the Innu and the LIA.

The Innu Nation has about 1700 people clustered in two Labrador communities: Davis Inlet and Sheshatshiu. The Innu are First Nations people, part of the great Algonkian language group that once occupied much of eastern Canada. Their traditional territory stretched across southern Labrador from the coast inland toward what is now Québec. Davis Inlet, the most northerly Innu settlement, is just 60 kilometres south of Voisey's Bay.

The LIA represents the Inuit, whose language is Inuktitut. The Labrador Inuit are among the most southerly of the Inuit nations, whose territory includes most of the Canadian Arctic.

The Innu are members of the Assembly of First Nations, the Inuit belong to the Inuit Tapirisat of Canada or ITC, which represents all of Canada's Inuit people. "Basically the Innu are more of an inland people, whereas we are mostly a coastal people," explains Fran Williams, the executive director of the OkâlaKatigêt Communications Society, which operates a network of Inuktitut radio stations broadcasting up and down the Labrador Coast.

Historically there was little love lost between the Inuit and Innu, and to this day there is little intermarriage or even intercoastal travel between the two groups. "I know there's one place called Massacre Island, though just who massacred whom I'm not really sure," Williams says with a rueful smile.

The Innu are widely regarded as the more militant of the two claimants to Voisey's Bay. The Innu Nation won widespread coverage and sympathy for their courageous, if unsuccessful, struggle against low-level flying from the huge NATO air force base at Goose Bay. In February 1995 a group of Innu from Davis Inlet rode snowmobiles north to the Voisey's Bay exploration site, surrounded the camp, and began a tense, three-week standoff that received scant news coverage. A force of 40-odd Mounties were called in to protect the 20 or so diamond drillers on the property, and the Mounties were themselves surrounded by some 80 Innu. Although there were rumours of some property damage, the incident ended peacefully. But it's the kind of thing that no one wants to see repeated and that lends urgency to the multi-sided negotiations now underway over the future of Voisey's Bay.

■

Their historic enmities notwithstanding, the Inuit seemed respectful of the Innu and their overlapping claim to Voisey's Bay in a way that few of them seemed to feel toward the Nunavik Inuit. At this point two parallel sets of negotiations were going on, with the Innu and the Inuit, and either of them could affect the future of Inco's operations in Voisey's Bay. The first phase of the negotiations involves land claims. The Innu and Inuit are each bargaining separately with negotiators from the federal and provincial governments. Both groups already had claims to their traditional lands pending for decades, but the Voisey's Bay mineral discovery has lent a sudden urgency to the matter. At the behest of Newfoundland Premier Brian Tobin all of the parties to both negotiations agreed to fast-track the talks, with a deadline of March 31, 1997.

The Inuit talks are said to be more advanced, with the Innu having signed a framework agreement for land claims (an agreement delineating the terms to be negotiated) only in March. The land claims talks, which are being held in St. John's, involve separate bargaining teams for the federal and provincial governments with the Innu and Inuit respectively.

A second, parallel series of talks is also underway with the Inuit and Innu, but these involve only the Voisey's Bay Nickel Company and are aimed at producing an Impact and Benefits Agreement (IBA) with both Aboriginal groups. These IBAs are taken exceedingly seriously in the Aboriginal communities because they are expected to spell out the terms of Aboriginal hiring and local purchasing, training provisions, environmental safeguards and Aboriginal input, and plans for de-commissioning once the Voisey's Bay deposit is mined out. The IBAs, which might be loosely compared to Collective Bargaining Agreements with Inco's Canadian unions, could end up costing the company many millions, and boost the cost of producing Voisey's ore accordingly. So how can it be said that, rich as the Voisey's ore undoubtedly is, nickel production will be virtually "free," paid for by copper, cobalt and gold production, when no one yet knows what the IBAs are going to cost?

I put that question to Mike O'Rourke of VBNC. "We just looked at what other such agreements, already in place, have cost," the affable O'Rourke shrugged. And those land claims negotiations? Doesn't the prospect of a breakdown, and consequent injunctions against mining until they are settled, cause the teensiest bit of anxiety in the Inco boardroom?

There are three categories of land claims agreements, explained O'Rourke, who spent fifteen years in the Northwest Territories negotiating just such agreements. There can be an award of surface and sub-surface rights, an award of surface rights only, or an award of neither but an acknowledgment that an Aboriginal claimant has the right to input regarding land use and environmental impacts. In the event that the Voisey's Bay claim wound up in court, Inco would mine the ore and pay potential royalties owing into a trust fund where the money would remain, pending a decision from the Supreme Court. And the arrival of the Nunavik Inuit, raising the spectre of yet another claim on the ground for which Inco had already paid so many billions?

"It's not our problem," O'Rourke said firmly. "That's up to the Inuit to decide among themselves."

Oh, it all sounds good, but as the old saying has it, "There's many a slip twixt the cup and the lip." And the space between the cup and the lip in the case of Voisey's Bay could prove a vast distance, indeed.

6

The View from St. John's

PERHAPS NOTHING so reflects the shifting fortunes of Sudbury versus Newfoundland more than the contrast between our respective provincial ministries of mining. Newfoundland's Department of Mines and Energy had just moved into spanking new quarters in a seven-storey building in Prince Phillips Square when I came to call. At the same time, Sudbury's Willet Miller Centre, a gleaming multi-million-dollar ministry complex at Laurentian University, was more than half empty and reputed to be on the auction block for a buck.

The difference, of course, is Voisey's Bay and the staking rush it has touched off both in Labrador and in Newfoundland. You can literally smell the fresh paint in the Natural Resources building as you await the elevator to the seventh-floor offices of Dr. Rex Gibbons, Newfoundland's Minister of Mines and Energy.

The man's credentials are just as impressive as his surroundings. A native Newfoundlander, Gibbons earned undergraduate and graduate degrees in geology from Memorial in St. John's before

A TRIP TO VOISEY'S BAY

graduating with a doctorate from the prestigious California Institute of Technology in 1974.

After spending two years working for NASA's Geology and Geophysics Branch at the Lyndon Johnson Space Centre in Houston, Texas, Gibbons returned to Newfoundland in 1976. He's been with the Ministry of Natural Resources/Mines and Energy since 1989.

I can't remember the last time I met a cabinet minister so eminently qualified for his portfolio as Dr. Gibbons, yet he remains down-to-earth and personable. I began our interview by asking Gibbons how much sulphur the Voisey's ore contained, because I knew that sulphur was the environmental wild card in the Voisey's Bay deck: just how much sulphur is in the ore? and just how would it be removed to avoid the kind of eco-catastrophe that happened in Sudbury earlier in the century?

The former geologist didn't have a hard and fast answer. "It's a sulphide ore body, like Sudbury," he frowned, so the sulphur content, presumably, would be roughly comparable. "We've got our concerns," he continued. "We have to meet the most modern environmental standards, and we will. It's going to be done by the year 2000 standards."

But what about an interprovincial agreement with Ottawa capping sulphur dioxide emissions at a fixed amount to reduce acid rain? (The agreement had been concluded in conjunction with Washington, which had secured similar standards on its side of the border.)

"Newfoundland is capped at 45,000 tonnes of sulphur dioxide (SO^2) per year," he confirmed, "and right now we don't even come close."

Not exactly, says Jim Brokenshire, spokesperson for a tiny, but vocal, minority of Newfoundlanders who are mightily concerned about the environmental impact of a new Inco smelter/refinery complex in the province.

"I was told by our Ministry of the Environment that we're at 44,500 tonnes right now," says Brokenshire, the mining caucus chair of the Environmental Mining Council of Newfoundland and Labrador, which is, in turn, a branch of the Canadian Environmental Network.

Brokenshire and I crunched some numbers. Assuming the sulphur content of the Voisey's ore is similar to Sudbury's, and assuming the new smelter will be state-of-the-art, like the Copper Cliff smelter, we calculated the annual SO^2 emissions. Sudbury's Inco smelter currently emits 800 tonnes of sulphur dioxide gas per day. Times 365 days a year, that equals 292,000 tonnes of SO^2, or more than six times as much sulphur dioxide as the entire Province of Newfoundland can legally emit right now.

A successful St. John's record store owner, Brokenshire had visited Sudbury a month earlier to see first-hand the environmental impacts of nickel smelting and refining. Whereas most Newfoundlanders are clamoring to have the smelter/refinery, and its jobs, located in their town or city, Brokenshire and his fellow environmentalists are more concerned about its impact on the fragile Newfoundland ecosystem. After spending a weekend in St. John's I could understand why. The area is similar, in many ways, to Sudbury. The terrain is rocky and soils are thin. The city, however, is not treeless. The streets are lined with maples, perhaps 30 or 40 feet high, which have been coaxed to maturity despite naturally acidic soils.

Brokenshire and others are convinced the smelter will be located somewhere in Placentia Bay, on the west coast of the Avalon Peninsula, the 'H' shaped finger of land on Newfoundland's southeast coast, on which St. John's is located. Many are betting that Placentia will be favored because it affords an ice-free deepwater harbour and is close to existing international shipping lanes.

St. John's environmentalists are concerned because Placentia Bay is just 100 kilometres upwind from their city. Given St. John's heavy annual precipitation, Brokenshire and others fear that, even if Inco's smelter fumes don't land directly on their city, they will create acid rain that will further imbalance already acidic soils. What's more, Inco's smelter/refinery complex will require huge amounts of hydroelectric power. The province's existing energy grid has no capacity to spare, and no access to either nuclear power generation or natural gas. And in our interview Minister Gibbons appeared to rule out hydroelectric sources. No, he repeated, Newfoundland Premier Brian Tobin's flap over the Churchill Falls contract with Québec had nothing to do with Inco's need for power.

To St. John's environmentalists, that spells just one thing: more thermal-generated (that is, oil-fired) power plants located somewhere near the smelter. Such plants are already Newfoundland's number one source of sulphur dioxide emissions.

Brokenshire et al. are hoping the Inco smelter will receive a full environmental impact assessment, but they also know the provincial economy is in desperate need of jobs; the depression caused by the collapse of the cod fishery is palpable in St. John's and the surrounding area. And they're afraid the economic and political momentum will outweigh their environmental concerns.

Indeed, just a few days before my interview with Gibbons, the province's Departments of Environment and Labour had announced that they were exempting Inco's proposed Voisey's Bay mine/mill operation in Northern Labrador from the province's Environmental Assessment Act, though it would be the subject of a joint federal-provincial environmental assessment panel.

But the fallout from Inco's smelter may have ramifications beyond St. John's and Newfoundland. The province will clearly have to seek an increase in its allowable SO^2 emissions from Ottawa, if our projections are correct. And here Newfoundland may find itself afoul of the Americans. Within hours of Bill Clinton's re-election he had dispatched Tim Wirth, his top environmental envoy, to Ottawa, according to Southam News columnist Giles Gherson. Wirth's message: cut greenhouse gas emissions like carbon dioxide and methane that contribute to global warming. So Washington was clearly zigging on air pollution issues, wanting tighter restrictions, while Inco, Newfoundland and Ottawa were trying to zag.

THE VIEW FROM ST. JOHN'S

Beyond a few people in the know, like Brokenshire, the full implications of a massive new nickel smelter in Newfoundland haven't really been felt yet, on the island or off it. But they will be felt, you can bet on that. And when they are, Inco's and Gibbons' confident predictions that Voisey's Bay ore will be smelted and refined in Newfoundland by the latter half of the year 2000 may yet prove wildly optimistic.

7

Irresistible Force, Immovable Object

THE IMMENSE PRESSURES of trying to bring the richest base metals find in Canada in a generation into production were clearly visible on Stewart Gendron's face. I ran into Gendron, the former Sudburian who had discouraged my attempts to reach the Voisey's Bay mine site six weeks earlier, in Thunder Bay at the November 1996 convention of the Canadian Aboriginal Mining Association (CAMA). Even though he was unhelpful in my journalistic enterprise, it was clear that the president of the Voisey's Bay Nickel Company had followed my coverage with interest. We spoke at considerable length, but our conversation was off the record, and there, I'm afraid, it must remain.

The CAMA conference also allowed me to meet Peter Penashue for the first time. Penashue is the president of the Innu Nation and the leader of some 1700 Innu people upon whose traditional lands Voisey's Bay is located. Both Penashue and Gendron, I discovered, stand a bit shorter than average height and both appear relatively youthful. The former appears to be in his 30's, the latter in his

40's. Penashue is a kindly, slightly stocky figure in horn-rimmed glasses under a thick thatch of gleaming black hair. Gendron, by contrast, was extremely pale and appeared on the verge of nervous exhaustion.

One of Gendron's biggest headaches just then must have been Penashue and his people. During a speech before some of Canada's most important mining executives Penashue excoriated Inco, warning, "We have made very little progress [in Impact and Benefit negotiations with Inco] in the last year, and without an IBA there will be no development." Mike Sopko and Stewart Gendron were both honorable men, Penashue concluded, "but are they Inco? Who is Inco? Inco has no human face." It was not easy to explain to his elders, Penashue continued, how it was that the investors in Diamond Fields Resources had made $4.3 billion at Voisey's Bay, and Robert Friedland had personally made $600 million in cash. "How can so many people make so much from Innu lands, while the Innu people make nothing?"

Nor was there much joy for Inco in the speech of William Barbour, president of the Labrador Inuit Association, the organization that represents the Inuit of Northern Labrador. "Inco's agenda will not be met unless the Inuit agenda is met," Barbour stressed. "There will be no development until the resolution of land claims." But in those talks, as well as in the LIA's IBA negotiations with Inco, "we remain in gridlock."

It was a pattern that would be repeated over and over around Voisey's Bay for months to follow: parties to closed door negotiations would seize on public occasions, and the news media, to slag one another in the most visible way possible. The behavior was baffling; were the parties using the media to posture and gain leverage in private? Could publicly traded insults be forgotten, and left at the entrance when closed door talks resumed? Or were these public outbursts a sign that some of the parties were acquiring a hearty dislike for one another? And if that were the case, how could they possibly combine to form genuine working partnerships over a

IRRESISTIBLE FORCE, IMMOVABLE OBJECT

multi-billion-dollar project expected to last over decades, if not several generations?

In discussions with Penashue over a traditional Aboriginal feast one night during the convention, I came to the conclusion that the Innu people do not want the Voisey's Bay nickel mine on their land under any circumstances, at least not in the foreseeable future. This position is in marked contrast to their Inuit neighbours. The Inuit, at least, have not given a flat "no" to the Inco development in the way the Innu appear to be doing.

Both Aboriginal nations were engaged in separate negotiations with the company over Impact and Benefit Agreements that will spell out the terms of the development vis-à-vis Aboriginal peoples. But, based on what Penashue told me, it's hard to see how Inco and the Innu can come to terms any time soon. "The timing is bad," is how Penashue put it.

I told him how envious the people of Sudbury were at his people's good fortune, how we would cut off our right arm to have another ore body like the one we've been mining for over a century. Could he not regard the Voisey's Bay ore body as a gift from the Creator? Penashue listened politely, nodded, and paused. "The timing is bad," he repeated, leaving me to ponder how it is that an impoverished First Nation of 1700 souls could walk away from the potential revenue, opportunity and employment of Voisey's Bay.

Without wanting to seem melodramatic, I can see the likelihood of an immense tragedy unfolding here, and it's personified by Gendron and Penashue. With the best will in the world, the biggest nickel producing company on the planet pays $4.3 billion for a newly discovered ore body, banking its corporate future heavily in the process. Simply put, Inco cannot afford not to develop Voisey's Bay, and to Gendron et al., delaying the development must seem as unthinkable as not developing it at all. But that's the clear message from both Penashue and Barbour at this conference: Voisey's Bay development must be postponed until their land claims are settled and a full environmental assessment can be completed.

A TRIP TO VOISEY'S BAY

Even then it's not at all clear that the Innu would allow development to proceed. But can they stop it, especially in the face of overwhelming support for the project among most Newfoundlanders and Labradorians? At the end of the day, probably not. But they can force Inco to pay a terrible price for its nickel. If the development proceeds over the objections of the Innu, Inco could face the sort of international campaign over Voisey's Bay that Hydro-Québec endured at the hands of the James Bay Cree.

Such a campaign could severely tarnish the image Inco has so carefully, and expensively, cultivated since the long strike in Sudbury in 1978-79: that of an environmentally—and people—friendly corporate empire. Or the Innu might mount legal action that could tie up the Voisey's Bay property in the courts for years, or even decades. Or, most unthinkable of all, the Innu might even organize armed resistance to the project in the homeland they call Nitassinan, conjuring up visions of another Oka. Now there's a way for Canada-First Nations relations to enter the new millennium. It's hard to see any way of avoiding some form of looming confrontation over Voisey's Bay. In Stewart Gendron and Peter Penashue, I felt I was seeing the irresistible force and the immovable object. Whatever happens, the nickel from Voisey's Bay, when it comes, if it comes, is going to come anything but cheap.

■

Stew Gendron and his superiors at Inco might have taken the cautionary speeches by Penashue and Barbour as a warning against the company's overly optimistic Voisey's Bay timetable. Instead, Gendron returned to St. John's from Thunder Bay and, within a week, called a news conference. The primary subject of the affair— the location of the smelter/refinery complex for Voisey's mill concentrate—was a matter of intense interest and speculation in Newfoundland and Labrador. Dozens of cities and towns in the province had vied for the massive project and the thousands of high-paying construction and industrial jobs that were certain to follow, notwithstanding the environmental impact that any metal smelting and refining operation were bound to entail.

As the St. John's environmentalists had both expected, and feared, the abandoned U.S. naval base at Argentia in Placentia Bay was the lucky winner, Gendron announced on November 26th, 1996. From the corporate point-of-view the fact that the huge former base, which had been built during the Second World War, had already been subjected to considerable industrial use was a huge plus. This made it a "brown field" as opposed to a "green field" site, with the unstated implication that environmental scrutiny might therefore be less rigorous.

But from the standpoint of the St. John's-based Mining Council, most of whose members resided in the Newfoundland capital, the choice of Argentia was ominous. Jim Brokenshire and his colleagues were quick to reiterate that Argentia was upwind of St. John's, and that any airborne pollution from the plant would quickly find its way to their city. "We're not opposed to a smelter and refinery somewhere in the province," Brokenshire stressed. "But we do question whether Argentia is the right place for it."

Nor was the environmental group reassured by the appearance of VBNC's Project Description Report on the Smelter/Refinery project, released ten days after Gendron's news conference. Inco attempted to meet the issue of St. John's air pollution head on, through assurances that the state-of-the-art technology that would be employed in the smelter would recover all but one percent of the sulphur released in the smelting process. "Thus discharges will be one-thirtieth of the Sudbury facilities, and less than many existing Newfoundland facilities," the report promised. If this promise was kept, the quick calculations Brokenshire and I had scribbled regarding the plant's SO^2 emissions would be vastly overstated.

Still, sulphur in some form, the environmentalists knew, is an inevitable by-product of any sulphide ore body. It is disseminated through the ore along with nickel, copper, cobalt and other more valuable elements, and it must be removed, in one way or another, during the milling-smelting process. In Inco's Sudbury operations, for much of this century the sulphur was simply burned off in the smelter, creating clouds of sulphur dioxide gas that both devastated the Sudbury environment and became a key ingredient in the global problem of acid rain. Brokenshire and other members of the Mining Council combed through the company's Project Description and soon discovered Inco's plans for the pesky sulphur: it would be "oxidized to sulphur dioxide and then converted to sulphuric acid" that would in turn be moved off the island via ship.

The largest single commodity produced by the Argentia facility, by far, Brokenshire noted with mounting alarm, would, in fact, be sulphuric acid. At a projected output of some 840,000 tonnes annually, according to Inco's figures, Argentia's sulphuric acid output would be nearly three times the amount of nickel, copper, and cobalt products combined, at some 300,000 tonnes annually. A lot of sulphuric acid would be passing through the shipping lanes off Southern Newfoundland.

The fact that Argentia was ice-free year-round was another of its chief appeals for Inco, which projected some 120 ship movements per year in and out of the smelter-refinery complex, or one arrival or departure every three days. But the environmentalists were concerned about another aspect of the Placentia Bay shipping lanes, which, as native Newfoundlanders, they knew all too well: while Argentia might be ice-free year-round, it was also notorious as the foggiest place on earth. Indeed, as the company's own Project Description report revealed, Placentia Bay averaged 206 foggy days annually. Fog-free days were, on average, highest in February, with sixteen, compared with just five fog-free days in July.

What were the environmental hazards of shipping all manner of industrial materials, including huge quantities of sulphuric acid, to and from the foggiest seaport on earth? It was one of many questions about the smelter/refinery complex that, for Brokenshire at least, cried out for answers.

IRRESISTIBLE FORCE, IMMOVABLE OBJECT

In the Placentia Bay area itself, however, the reaction to Gendron's announcement was ecstatic. Like most of the rest of Newfoundland, the region's economy had been hard hit by the collapse of the cod fishery and the resulting cod moratorium. The Inco decision promised sure-fire economic deliverance. The $1.05 billion project would require some 3000 workers during the peak construction phase, according to the Description Report, an extensive document packed with tantalizing details, including a list of construction contracts that would be let prior to the beginning of construction. The tone of the report was so definite, so confident. "Further design and feasibility studies are underway such that the project is anticipated to be operational by the year 2000," the Preface stated. "In 1995, Inco produced 400 million pounds of nickel and purchased an additional 200 million pounds to supply its customers. Currently there is no excess world nickel smelting capacity that can process Voisey's Bay concentrate at the expected production rate of 270 million pounds of nickel per year," the report continued. "For Inco's shareholders, the Voisey's Bay project will satisfy its corporate objective of growing with the nickel market and allow the substitution of Voisey's Bay nickel for purchased nickel. Inco has targeted a production level of 750 million pounds of nickel per year by the end of the year 2000."

What was there not to like? It was clearly a win-win situation. In return for a sizeable capital investment in Argentia, Inco would nearly double its nickel-producing capacity in the span of four short years. Besides the construction boom, the residents of the Placentia Bay area would benefit from the creation of a thousand high-paying, long-term industrial jobs at the complex, plus at least as many new jobs that would result from the economic spin-off. The news touched off a mini-land boom in the area, as speculators rushed to buy property that would, they were sure, appreciate in the coming months. Modest, grassroots economic development projects were soon abandoned in all the excitement.

Inco's detailed plans for Labrador and Newfoundland galvanized the business communities of Goose Bay, Argentia, and even

Sudbury. Over the winter of 1996-97 entrepreneurs from all three places, formed into official trade delegations with the encouragement of Inco and VBNC, exchanged visits and official tours to facilitate "partnerships" between Northern Ontario and Newfoundland and Labrador.

An enterprising Liberal MP, Ray Bonin of the Nickel Belt riding near Sudbury, organized one such exchange, and dubbed his band of Sudbury businesspeople "Team Sudbury," an evident homage to the annual Team Canada business junkets led by Jean Chrétien to Asia and Latin America. Coincidentally or not, the Team Sudbury mission took place in the run-up to the June 1997 federal election. Bonin was re-elected handily.

Reporters in Sudbury, St. John's and Goose Bay gave these exchanges copious coverage in stories that were invariably brimming with business confidence in the Inco-fed cornucopia that awaited just around the corner. The myriad of thorny issues that hung over the project—unresolved Aboriginal land claims and Impact and Benefits Agreements, the environmental assessment process, the source of hydroelectricity to fuel the smelter and refinery, the tax regime that would apply to this apparent bonanza—was quickly brushed aside, if addressed at all. Confidence was high, expectations were boundless. They were heady days.

A Season of Discontent

Summer 1997

8

Inco Drills a Bootleg

BY THE SPRING of 1997 an increasing air of unreality had settled over the development of Voisey's Bay, and Inco's annual shareholders' meeting, held in late April in Toronto, did little to dispel that impression. Here are some of the key assertions, made by CEO Mike Sopko and stated baldly as fact, before the company's investors, lenders and the financial press.

- Inco will produce mill concentrate from Voisey's by the end of 1999 and finished nickel from the smelter/refinery complex by the end of 2000.
- Inco will proceed to install infrastructure—a road, airstrip and wharf—at the Voisey's Bay mine-mill site this summer, 1997.
- The company has drastically increased the anticipated size of its mine-mill workforce in Northern Labrador, once the Ovoid open pit is mined out, seven years after mining operations begin.
- Company environmental impact studies will be completed by the end of June and final environmental approvals will be obtained by May 1998, according to Stewart Gendron.

Six months before the meeting, after returning from my own fact-finding tour of Labrador and Newfoundland, I had reported that Inco's timelines, which Sopko was still repeating, were "wildly optimistic." Yet the company continued to cling to predictions that became more untenable by the day.

Take the thorny issue of land claims, for instance. In the fall of 1996, it was announced that land claims negotiations, involving the Labrador Inuit Association and the Newfoundland and federal governments, were to be "fast-tracked" in an effort to reach an agreement by March 31, 1997. That deadline had come and gone and, instead of having an agreement that might clear the way for Voisey's Bay development, negotiations with the Inuit had broken off—indefinitely.

As always, I found Inco's up-beat Bay Street pronouncements and the reality on the ground in Labrador to be so wildly divergent that I had to wonder, not for the first time, if Mike Sopko was referring to the same Voisey's Bay as the one I visited.

Just for fun I decided to call the Labrador offices of the Innu Nation the morning after the Inco annual meet, to learn their reaction to the pronunciamentos of Messrs. Sopko and Gendron. There was, first of all, the simple logistical problem of completing a phone call. Every number I dialed in Northern Labrador, including Inco's own community office in Nain, produced a busy signal, causing me to wonder if all the telephone lines were down, or otherwise dysfunctional.

"Yes, the phone lines were out here this morning," apologized Christine Cleghorn, Voisey's Bay Assessment Co-ordinator for the Innu Nation, when we finally connected. "The weather here has been really bad the last week—fog and blizzards—and sometimes the phone lines go out, too."

I well remembered the Labrador weather, especially the coastal fogs so thick even helicopters couldn't fly, sometimes for days at a time. Inco was now saying it plans to move mill concentrate by ship nine months of the year, fog or no fog, and through pack ice, to the even foggier smelter in Argentia. And to fly the workers in and out of Voisey's on a regular timetable.

"There's no way," Cleghorn responded when I asked if she believed Inco would really have nickel concentrate out of the mine site by the end of 1999. "They blew that deadline a long time ago," she declared flatly.

Was she aware that, in a few short weeks, Inco planned to begin work on the airstrip, wharf and road essential to any further development at Voisey's Bay? "We haven't been told a thing, although with the bad weather communications can get a little slow around here.... They can't go ahead without permits, and they don't have them. We fought them last year and won, and if we have to, we'll fight them again this year." The situation is a little like a developer coming in to throw up a 10-storey apartment building in your backyard without notification, planning approvals, a building permit, or even clear title to the land.

Cleghorn was equally dubious about Gendron's optimistic timetable for environmental approvals. The week before, it turned out, an environmental "scoping" panel was touring Labrador communities, holding hearings aimed at getting the terms of reference for Inco's Environmental Impact Study, or EIS. Even though that panel's work was not yet complete (the panel was weathered in on the coast for several days), Gendron told the Toronto press that the company's EIS, the parameters of which had not yet been established, would be completed by the end of June, or just two months hence.

Interested parties would then have 75 days to review the EIS, and a second, and far more rigorous, round of environmental hearings would begin, according to Cleghorn. "And how much credibility will the EIS have if it's thrown together in two months?"

Cleghorn was as intrigued as I was at Inco's announcement that "the workforce requirements for its planned mining and milling operations at Voisey's Bay could be expected to increase significantly once mining operations move underground after approximately the first seven years of operation . . ." The workforce might now range from 1500 to 2000 people, a trebling of the original estimates of 600.

That announcement was made on the same day that Inco president Scott Hand, Mike Sopko's Number Two, was publicly mulling over the probability that the company would have its first "minerless mine" up and running in Sudbury by the year 2000. This will mean further workforce downsizing in Sudbury, and greater productivity, according to Hand. So Inco can run a "lights out" operation (will the last worker out of the plant please turn off the lights?) in Sudbury, but needs thousands of workers to mine the Voisey's Bay ore body? And what will that mean to the cost of a pound of Voisey's ore versus a pound of Sudbury nickel?

Meanwhile, the lack of land claims settlements with the Inuit and Innu were beginning to take a toll on Inco's public and corporate image. Shareholders at the annual meeting were greeted by a picket line organized by a new group calling itself the Voisey's Bay/Innu Rights Coalition.

Founded in January 1997, the Toronto-based Coalition includes supporters from the Anglican Church, the Canadian Environmental Defence Fund, the Citizens for Public Justice, the Toronto Catholic Worker's Movement, the Christian Reformed Church, the Friends of the Lubicon, the Mennonite Central Committee, and the Voice of Women for Peace, among others, co-ordinator Lorraine Land explained.

"We took a fairly low-key approach at the Inco meeting, but we're expecting to heat up from here. If there's no breakthrough on land claims in Labrador we expect a more volatile situation," Land warned.

Inco may think it can bull ahead with infrastructure building and mines and mill development before coming to terms with the Inuit and Innu, and technically and logistically it probably can. But politically? Morally? Legally? I'm not so sure.

"They see the Innu and Inuit as specks of dust to be brushed aside on the way to development," Christine Cleghorn noted ruefully. But she knows what Sopko and Gendron don't yet seem to

INCO DRILLS A BOOTLEG

fully appreciate: these specks of dust are people, armed with all of the weapons of the Information Age, and with a coalescing base of support in urban Canada.

These are people, moreover, who had already proven that they could be expected to stand tall and proud in defense of their rights and homeland. And the ore Inco wants so badly to exploit is right in their own backyard. As the events of the next few months were about to show, the struggle over Voisey's Bay wasn't over yet. In fact, it had barely begun.

■

Sooner or later, hardrock mining comes down to precisely this: breaking ground, often through some of the most ancient, obdurate rock on the crust of our planet. For most of the 20th century this was accomplished in a laborious fashion, using a succession of bulky, hand-operated mechanical drills. The miner working at the face, or heading, advanced at the rate of perhaps eight feet per eight-hour shift by drilling a series of holes in a roughly circular pattern. The drill holes were then filled with high explosive and wired together with igniter cord and blasting caps so that, within milliseconds, the rounds would fire in a carefully predetermined sequence.

If he was lucky, the miner would return to his stope following a blast to find a "clean break"—and a face, or breast, that was more or less sheer. But if the ground was faulty, or the shots were incorrectly coordinated, or if one of them had misfired, the unlucky miner would find part of his drill hole still evident in the face. This sinister hole was known, in miner's parlance, as a bootleg. A clear procedure was called for in the event of a bootleg: the hole was to be carefully flushed out with water, so that no unexploded powder would remain. But an incautious miner, greedy for bonus and therefore loathe to spend his time on such an unproductive pastime, might decide to ignore this precaution and simply begin drilling the holes for his next round, a process known as "drilling the bootleg."

Drilling the bootleg was a foolhardy venture on several counts: the lingering powder might explode in the miner's face, ignited by friction or a spark from the drill steel, with lethal results; and, because it was so risky, drilling the bootleg was strictly frowned upon by mine management. If a shift boss happened along, it was grounds for immediate suspension without pay, if not outright dismissal.

Yet drilling the bootleg was precisely what Inco's senior management elected to do in the political powder keg known as Voisey's Bay in the spring of 1997. The immediate issue was the decision to press ahead, over the objections of both the Innu and the Inuit, with infrastructure construction at the mine/mill site. This decision, announced by Mike Sopko at the annual shareholders' meeting in Toronto, would have long-term consequences for Inco that would prove as disastrous as they would be unforseen. It was, in hindsight, perhaps the single most damaging decision the company would make in its long struggle to bring Voisey's Bay into production.

In all the contentious issues that surrounded the development, there was, by the spring of 1997, a single bright spot: the Memorandum of Understanding, or MOU, concerning the environmental assessment procedure that would be followed for the mine/mill project. The MOU, signed on January 30th, 1997, was the only agreement that had been successfully negotiated among the principal parties. The federal government, the government of Newfoundland and Labrador, the Labrador Inuit Association, and the Innu Nation had each agreed to cooperate, and to "streamline" the procedure by collapsing provincial and federal environmental reviews into a single environmental assessment panel.

Whether the infrastructure work—construction of an airstrip and a road linking Discovery Hill with the main camp at Anaktalak Bay—would fall within the purview of the MOU had been discussed in negotiations preceding the signing of the Memorandum, and the LIA and Innu Nation both believed that it would.

But through a semantic sleight-of-hand, by terming it "Exploration Support Work," as opposed to permanent infrastructure leading directly to the further development of the project, senior Inco management in Toronto clearly hoped to sidestep the environmental procedure and complete the construction, so crucial to permanent development, during the summer of 1997.

One of the immediate consequences of this decision was a setback in the relationship between Inco and both the Inuit and Innu peoples of Northern Labrador. Again and again senior executives from the head office of the Voisey's Bay Nickel Company in St. John's had assured leaders of the LIA and Innu that the infrastructure work would not proceed without their consent. Indeed, Herb Clarke, one of six vice-presidents ensconced in VBNC's posh headquarters in downtown St. John's, had given his assurance to the Innu on audio tape. By reneging on their undertaking, Sopko and the other senior officers at Inco headquarters in Toronto had hung their men on the ground in Newfoundland and Labrador out to dry. In the case of Clarke, who was charged with negotiating Impact and Benefit Agreements with the Innu and Inuit, a delicate task requiring both diplomacy and the establishment of a considerable degree of trust, the decision by his superiors would prove particularly egregious.

But, like the foolhardy miner greedy for muck, Sopko and the others had compelling reasons for their desperate gamble. World nickel prices had begun to fall, and with them, the company's profitability. Share prices were heading south, too, spurred, in part, by investor impatience at the slow pace of development in Voisey's Bay. And, overhanging everything, was the huge investment Inco had already made in the project, with very little to show for it.

On May 22nd Inco applied to Newfoundland's Ministry of Environment and Labour for the requisite permits to begin work on its "advanced exploratory infrastructure." Although the scale of both the road and the airstrip had been reduced in its application, the move elicited immediate protest from both the Innu and the

Inuit. On June 16th, the Environment Assessment Panel itself joined the fray. In a letter addressed to Kevin Aylward, Minister of Environment and Labour for Newfoundland and Labrador and to Christine Stewart, federal Minister of the Environment, Panel chair Lesley Griffiths "on behalf of the Panel," cautioned the ministers about the consequences of granting the infrastructure permit.

". . . The Panel is concerned that approval of Exploration Support Works could jeopardize or delay the review process defined by the Memorandum of Understanding (MOU)," Griffith warned. "If approval were given to the Exploration Support Works, the Panel believes that the credibility of the review process would be called into question and that some communities or individuals would discontinue their participation. This would obviously damage the Panel's ability to carry out an effective and timely review of the Undertaking."

Despite this, Aylward's ministry granted Inco's application on July 2nd, ruling that the work in question was "a separate undertaking from the mine/mill development undertaking that was exempted from the Newfoundland Environmental Assessment Act in favour of the MOU." As Inco hastily made plans to ship heavy earthmoving equipment north by barge and the buzz of chainsaws began to be heard through the modest, but primeval, spruce forests of Voisey's Bay, the Innu promptly filed for an injunction to stop the work before the Trial Division of the Newfoundland Supreme Court in St. John's.

The joint Innu-Inuit motion was filed on July 8th, and the Innu Nation issued a press release on the same day. "It's too bad we had to go to court," observed Innu Nation president Katie Rich, who had defeated Peter Penashue's re-election bid in January 1997. "This action is about forcing VBNC to live up to its public statements about its commitment to the environment. The company can't claim to be environmentally responsible when it is trying to undermine the environmental review process established under the MOU."

Just ten days later, on July 18th, the Court denied the injunction application and upheld the Newfoundland government's decision to issue the work permits. The Exploration Support Work was "necessary to support the ongoing exploration activity," the Court ruled.

Another spate of angry press releases and another flurry of legal activity followed, as lawyers for the Innu and Inuit prepared to appeal the decision to the Appeals Division of the Newfoundland Supreme Court. "We are reviewing all of our options now. You can be sure that the Innu will do whatever is necessary to protect our rights!" vowed Rich.

Privately, Rich and her counterparts at the LIA were not optimistic about winning further court battles at the provincial level. Brian Tobin's government was a keen supporter of the Voisey's Bay development, as evidenced by the granting of the work permits, and the chances that judges prominent in St. John's tightly knit business and legal community would uphold the arguments of a group of Labrador Aboriginals appeared slim.

But then, in late August, the phone in Katie Rich's office rang, even as the foolish miner, heedless of the bootleg, collared his hole, took a deep breath, and began to drill. . . .

9

Seven Days at Voisey's Bay

THE WAY KATIE RICH told me the story six months later, the whole thing started with a phone call from Chesley Anderson, the mineral resource advisor for the Labrador Inuit Association. The LIA's Impact and Benefits Agreement negotiations with Inco over the Voisey's Bay nickel property were flagging and, worse still, the company had begun infrastructure work on the site against the express wishes of both the LIA and Rich's own Innu Nation. After holding several community meetings in Nain the LIA had decided to occupy the site and, if possible, physically disrupt the construction work. Would the Innu be interested in participating? Rich said they might, but she'd have to consult with the Innu community and elders first. The Innu are old hands at confrontation and civil disobedience—they'd occupied the Voisey's Bay project once before, in February 1995—but such tactics are relatively new to the Inuit.

"Who's your protest coordinator?" Anderson asked, a query that still makes Rich smile, because it strikes her as, well, such an *Inuit* kind of question.

"We don't have one," Rich remembers telling Anderson, and her smile becomes a laugh. "Things just happen."

And so, a few days later began the great Inuit-Innu occupation of Voisey's Bay, an action that would, in tandem with a pair of signal court decisions in far-off St. John's, prove disastrous, perhaps even fatal, for the largest nickel mining company in the world.

"Was it fun?" I ask Rich.

The 38-year-old president of the Innu Nation smiles again at the recollection of those seven days at Voisey's Bay. "It *was* fun, yeah."

■

They are known as the Mushuau Band of the Innu Nation, which means, literally, "The People of the Barrens," but I think of them now as "The People of the Caribou," because of their almost mystical connection with the great George River caribou herd, or "People of the Mist," because of their uncanny ability to materialize, as if out of thin air, almost anywhere and any time on Nitassinan, as they call their homeland. Their peregrinations are legendary, and have often been described as "heroic" by white missionaries, fur traders, and anthropologists. For thousands of years before the coming of Europeans the Innu—women, children, men and elders—travelled thousands of kilometres on foot each year across the achingly beautiful yet frequently lethal terrain of central Labrador, surviving entirely off the land, or *nutshimit*.

Together with the Sheshatshiu Band, based 280 kilometres to the south near Goose Bay, they constitute the 1700 members of the Innu Nation, and they are, according to Memorial University anthropologist Adrian Tanner, arguably the most culturally and materially intact First Nation left in North America. Owing partly to the remoteness and inhospitability of Nitassinan, Innu contact with Europeans, when it finally came, was both late and disastrous. It began, so far as anyone now knows, with something called "The Ungava Venture," in about 1830, Tanner explains. Perhaps attempting to forestall suspected competition for the fur trade from Moravian missionaries, who had begun colonizing the Inuit of the northeast Labrador coast as early as the 1770's, the Hudson's Bay Company established posts on the western side of the Ungava Peninsula to trade with the Innu. After making the latter dependent on European trade goods, notably rifles and cartridges, the HBC unaccountably abandoned the Ungava Venture, causing widespread starvation and death amongst the Innu.

The Innu returned to the mists of Nitassinan and were more or less forgotten by European interests for decades. Around the turn of the century the George River caribou herd went into a steep decline, for reasons that remain obscure. This wreaked considerable hardship on the Innu, who were dependent on the caribou for food, clothing, and even shelter, since their tents were covered with caribou hides. The shortage of caribou led directly to a geopolitical division of the Mushuau Band, with one half settling on the Québec side of the Québec-Labrador peninsula, and the other settling, eventually, in what is now Davis Inlet. The former, now known as the Naskapi of Schefferville, and the Montagnais people of Québec are all Innu, though they are not considered part of the Innu Nation of Labrador.

As the twentieth century wore on the Mushuau Innu began to have increasing contact with Roman Catholic priests, who encouraged the people to abandon their traditional lives in *nutshimit* and move into a settlement where children could be sent to school, the community could attend mass regularly, and European health care could be administered most conveniently. The people complied, and were moved to a succession of locations, generally without consultation or consent. "The Innu were treated like cattle," is the way Katie Rich describes it. The most disastrous relocation of all occurred in the 1950's when federal bureaucrats shipped the Innu to Nutak, a coastal settlement north of Nain. They apparently failed to realize that the location was deep inside Inuit territory and at least 100 kilometres north of the northerly border of Nitassinan, which happens to fall roughly at Voisey's Bay.

The Mushuau Band made another of their heroic marches, and the entire community retreated on foot across the ice to the friendlier confines of Nitassinan. Eventually they were relocated to the current Davis Inlet site, a place most people in the community have long since come to despise, partly because it is on an island, making travel impossible during fall freeze-up and spring break-up, partly because an adequate supply of fresh water has never been found (most houses in Utshimassits remain without running water), and partly because, once again, they were not consulted in the selection of the settlement's location. Utshimassits means "Place of the Boss."

By the early 1990's Davis Inlet had become synonymous with absolute despair. Gas-sniffing among the youth and alcoholism among the adults were rampant, and on February 14, 1992, the community hit rock bottom, when a house fire claimed the lives of six children. Adults and elders stood helplessly by as the dwelling burned, because there was no water to extinguish the flames. "Confused about whether or not there was anyone in the house, men went from one house to another to look for the children," Katie Rich wrote three years later about that terrible day. "Daylight came, and we saw the bones. All day elders, women and children came and examined the ashes, stood around in the freezing cold, and cried." There is a high, rocky spire overlooking Utshimassits. I'm told it's known locally as "Satan's Hill," because Davis Inlet has had more than its share of attention from the Devil.

■

The first Inuit protestors arrived at Inco's Voisey Bay camp at Anaktalak Bay around mid-day on Wednesday, August 20th, 1997. They arrived aboard a steel tug, aptly named *The Checkmate*, which is owned by the Labrador Inuit Development Corporation. An all terrain vehicle and a rowboat were used to ferry the occupiers, who would become known as "The First Wave," onto the beach.

The LIA stalwarts decided to pitch their tents some distance from the Inco camp and up a steep hill, smack in the middle of the company's road work. As huge backhoes, or "high hoes," and earthmoving trucks roared around them, the Inuit calmly pitched their tents and settled in for an indefinite stay. The weather was warm and sunny, but insects were a ubiquitous and constant torment. Centuries of survival in an environment that is, if anything, even more inhospitable than the Innu's have made the Inuit a prudent and patient folk, and they began their occupation with almost military precision. They were equipped with a generator, freezer, stove, and satellite telephones to maintain constant contact with the outside world.

At first Inco's contractors tried to ignore the small tent village that had suddenly sprung up in their midst, but on Wednesday night it must have become apparent that the Inuit had effectively surrounded millions of dollars' worth of heavy equipment that could become an inviting target of vandalism. The issue was decided just after sunrise on Thursday morning, when construction workers about to start the day shift found a makeshift barricade blocking their path.

As Inco security personnel looked on with stoney faces, the huge backhoes and dump trucks were formed into a convoy and, amidst much squeaking of bulldozer tracks and roaring of diesel engines, they lumbered their way down the hill and back to the security of the main Inco camp at Anaktalak. Round one had gone, and quickly, to the Inuit; the controversial road work had been stopped. No one knew that morning that it would be a long, long time before it resumed.

■

The Innu penchant for direct political action that would inspire the occupation of Voisey's Bay and help in the spiritual revival at Utshimassits actually began in the sister community of Sheshatshiu in March 1987, when a group of hunters openly defied a provincially imposed ban on caribou hunting in the nearby Mealy Mountains. On March 15th an RCMP raiding party swooped into the hunters' camp aboard a helicopter. Arrests were made and fines were eventually levied by a provincial court judge in Goose Bay. "Although the Mealy Mountain protest did not cause the Newfoundland government to change the [hunting] legislation, it did accomplish something more important: it encouraged the Innu to keep fighting back until their rights are recognized," writes journalist Marie Wadden in her book *Nitassinan: The Innu Struggle to Reclaim Their Homeland*.

The Innu had resumed their strategy of civil disobedience, this time in a far more spectacular fashion, in the fall of 1987. Their target was the joint NATO low-level flying program operated across Nitassinan from the sprawling airfield at Goose Bay. At first the Innu occupied parts of the bombing range, forcing the delay or cancellation of planned flights, but, on September 15, 1987, the women of Sheshatshiu broke into the airbase and led a sit-down protest right on the runway. It was a courageous, even audacious thing to do. Not only were the 75 or so women defying the power of the Canadian state and military with their bodies, they were also confronting the most sophisticated aerial weapons of war ever built.

A few Innu leaders were arrested and the rest withdrew voluntarily, but a second, and larger, runway protest was held a week later. Despite arrests and jail sentences that took the mothers of Sheshatshiu away from their children while the former were incarcerated in Newfoundland, there would be seventeen Innu incursions onto the base, ten onto the runway, over the next two years.

In media interviews and statements from the prisoner's dock, the Innu made the same point over and over again: Nitassinan was their homeland. They had never signed a treaty with the Crown (and neither, for that matter, had the LIA), the land had belonged to them, by use and by occupation for millennia, and provincial hunting regulations, low-level aerial bombing, and even the presence of the Goose Bay base itself was a direct violation of Innu sovereignty.

Although the protests failed to stop the low-level flying program, the resistance galvanized the Innu as a people. When the federal government refused to appoint a public inquiry following the Utshimassits house fire in 1992, the Mushuau band, led by then Chief Katie Rich, decided to hold a people's inquiry of its own. The inquiry was called "Gathering Voices: Finding Strength to Help Our Children." Its findings, and much of its testimony, were published as a bilingual *Innu-eimun* and English book in 1995. At a series of public and private meetings people of all ages spoke candidly, and for the first time, about the twin plagues of solvent sniffing and alcoholism. "Gathering Voices" was the beginning of a long road to healing for the people of Utshimassits, but so too was the direct political action that flowed from it.

■

The three days that followed the retreat of the road-building crew were largely uneventful, marked mainly by the build-up of forces on both sides. The Innu began to arrive, in ever greater numbers, from Utshimassits, from Sheshatshiu, and even from the Québec-Naskapi-Montagnais side, aboard boats and on a series of chartered float planes. Eventually there would be some 200 Innu on the site. But unlike the Inuit, the Innu brought the whole community, including children and elders. The Innu nonchalantly pitched their tents near the shore of Anaktalak, much nearer to the main Inco camp. The men set out to hunt porcupine for dinner in the surrounding bush, and the voices of children soon filled the pristine air. It was as if an entire village had suddenly just appeared out of nowhere, and its residents began their appointed tasks on *nutshimit*, much as they had always done. A young non-Aboriginal reporter from the OkâlaKatigêt Communications Society in Nain remembers feeling a sense of awe as he watched the Innu set up their camp. It was as if he had entered some kind of time warp. "Look, Paul," he blurted out to a colleague, *"real Indians!"*

But the build-up of RCMP officers continued apace, too. Their numbers approached 75. They were cordial, almost deferential, at least in their dealings with LIA president William Barbour, a video-taped record shows. But they were also heavily outnumbered by the Inuit and Innu combined, and, like Voisey's Bay itself, they too were surrounded.

The occupation received scant news coverage. CBC Radio in Goose Bay sent a reporter, and there were one or two others, but virtually all of them, except for a video crew from the OkâlaKatigêt Society, left the site when the first wave from Nain returned home, as planned, on the morning of Sunday, August 24th. Which was a shame, because after their departure the real show began. The Innu had decided that it was time for something, in Katie Rich's words, to "just happen."

■

For all its hype as the "base metal discovery of the century" in the Bay Street press, the main camp of the Voisey's Bay Nickel Company at Anaktalak Bay isn't much to look at. At its heart is a series of low, one-story ATCO trailers huddled densely together. Even though they must enclose an impressive acreage, the office-living-dining quarters also convey a sense of both claustrophobia and impermanence, surrounded as they are by the immense grandeur of the North Labrador coast. On Sunday afternoon the Innu decided to go walkabout around, and into, the VBNC trailers. They clearly caught both the RCMP, and the company itself, completely unawares.

Katie Rich and a group of women barged into one trailer and promptly sat down on the floor. Outside the gravel patch around the trailers was suddenly alive with children, adults and elders, laughing and smiling, wearing mass-produced T-shirts that read "INNU land, not INCO land," and brandishing professionally made picket signs bearing the same message.

Mounties in coveralls spilled from the trailers to face the protestors, with their backs against the thin sheet-metal walls. Inside, meanwhile, more Mounties converged on the trailer where the women were conducting their sit-down. They wanted to arrest Katie Rich, but she was protected by a tight clutch of supporters. The confines were extremely close, almost airless, and the place was a bedlam. One of the women grabbed a Mountie's radio and handed it to Katie. She began to talk into it, speaking only *Innu-eimun*.

"She's got the radio, she's got the radio," Rich heard a male voice crackle back, in English. As more and more Mounties crowded into the room the floor of the trailer suddenly gave way under the weight of so many bodies. It was only a few feet to the ground below, so no one was injured, and Rich remembers feeling a sense of relief at the fresh air that suddenly came in through the gap that had opened underneath the trailer. The protestors, followed by the Mounties, clambered into a second trailer, and the same thing happened, and then into a third. The women were howling with laughter at the whole thing, as Inco's offices were trashed one by one.

Outside the kids were beginning to have fun, too. Some of the trailers have re-bar ladders welded to their sides, which the youngsters quickly scaled. They began to caper on top of the trailers, playing king of the mountain, waving picket signs. A couple of them casually pulled cables (electrical? satellite? telephone?) out of one trailer, while others beat a tattoo on the roof with their fists. Eventually a few Mounties clambered reluctantly up on to the roof, too. They were much bigger than their quarry, but they were also much slower, and older, and the Innu boys dodged delightedly out of the way. How the Mounties, who were flown in from hundreds, maybe thousands, of kilometres away must have felt to be playing this hopeless game of tag atop ATCO trailers in the godforsaken wilderness of Northern Labrador can only be imagined.

The adults and elders watching from the ground enjoyed the show immensely, grinning, laughing, shouting encouragement to the youngsters in *Innu-eimun*, until finally the elders told the kids to come down. They did. Eventually, Katie Rich and two other women were finally arrested, and tension quickly built amongst the onlookers outside. A couple of Innu leaders from Sheshatshiu were allowed in to talk with Katie. They came back out to reassure the crowd that everyone was all right, and that Rich had refused to sign her own arrest warrant. The Inco camp was probably spared when the Mounties decided to let Rich return to the crowd of demonstrators outside.

But the fun wasn't over, yet. The kids had discovered a pile of diamond drill core samples, stacked neatly in wooden trays five feet high outside the core shack. They scooped up handfuls of the precious, carefully arranged core, the only physical record Inco had of the ore body underfoot. The samples were smooth and round and with a fine heft and their utility was clear to any ten-year-old: the core makes perfect ammunition.

Suddenly, though, everyone's attention was drawn to a tiny, birdlike Innu woman who, in *Innu-eimun*, began to berate the Mounties guarding the door. Her name was Monique Rich, an elder from Utshimassits, and she is, as it happens, the mother of Katie Rich. She was born at Voisey's Bay, and her father died and is buried here. Inco salvage archaeologists have so far been unable to find his burial site. A handsome, thickly set Innu man wearing a red bandana and looking every inch a warrior began to translate the elder Rich's words into ringing, cogent English.

"She wants to see the bones of her father. You destroy this land. I will not quit fighting. I'm gonna build a fire here! Shut up, inspector, you shut up." The kids surrounding her loved this, their elder dissing the Mounties, and they laughed and howled into the faces of the police.

"I want to see my father's bones. Are you going to show me my father's bones? When I was here there was no white man. We weren't fed by white people. I still see my father's trails here." With sweeping gestures Monique Rich indicated the towering hills and broad waters of Anaktalak Bay, steely gray now under a leaden sky.

"If the chopper lands here again I'm going to start throwing rocks. I want to see my father's bones! This is our land. In all these shores used to be Innu camps. When did you guys camp out here a long time ago? Tell me! I'm waiting! Tell me!" demanded this tiny woman. A few of the kids repeated her words in *Innu-eimun*. The Mounties, utterly expressionless, had nothing to say.

"We never saw the RCMP here a long time ago. Why are you standing there now? Tell me! I'm waiting! This is where I was born . . ."

Perhaps as she intended, Monique Rich's peroration defused the mounting tension somewhat. Most of the core samples were never thrown, and more serious damage to the Inco camp was averted. But the Innu had sent the company a message loud and clear. This is their land, they can come here, without warning, any time they want, and Inco's development at Voisey's Bay is at their mercy. Without Innu permission for the project, life at the Voisey's Bay camp can be turned into pure, rattling hell. Indeed, a number of employees who spent Sunday afternoon huddled inside the trailers were reportedly so shaken they asked to be sent home the next day.

■

On Monday, August 25th the protestors received almost unbelievable news from their lawyers in St. John's. A provincial Supreme Court judge had granted their injunction and ordered a halt to all further construction work at the camp, pending the outcome of a joint LIA-Innu lawsuit that was yet to be heard by the courts. Their actions on the ground had received the imprimatur of law. Reasoning that continuation of construction might undercut the eventual outcome of the full LIA-Innu case, the judge in St. John's ruled "that, in this case, justice delayed could truly be justice denied." The protestors were jubilant, and they celebrated that evening. For once, the White Man's law was actually on their side.

On Tuesday, August 26th, the LIA people struck their tents and headed for home, but Katie Rich elected to linger for several more days to oversee the break-up of the Innu camp. She also accepted Inco's generous offer of its helicopters to ferry some of the elders home and, of course, to help get the Innu the Hell out of Dodge.

Seven days after it had all started, Katie Rich was the last of the protestors to leave the beach at Anaktalak. She wended her way contentedly back to Utshimassits by boat, past the rocky headlands and through the same bays and fjords and islands of her beloved Nitassinan that had been travelled by her mother, and by her mother's father, and by a clean, unbroken line of Innu beyond memory, and beyond all knowing.

10

The Bootleg Explodes

THE INNU-INUIT protest combined with the temporary court order enjoining Inco from continuing with its infrastructure construction was a serious setback for the company, but over a ten-day period in mid-September, worse was to come. Already, the precious days of Northern Labrador's short summer construction season were being lost to the company, and any hope of completing the "Exploratory Support Work" in 1997 was waning as fast as the daylight hours in the Canadian sub-Arctic. But fresh disasters were about to strike, and the summer of 1997 was about to end in still deeper discord for the Voisey's Bay project.

The season of discontent began to draw to a close on Friday, September 12th, when the Innu Nation walked out on its IBA talks with the company. "Voisey's Bay Nickel Company has once again demonstrated that it is not taking Aboriginal people seriously," Katie Rich noted in a press statement released three days later. "We came to the table in good faith, believing that we could reach an agreement in principle on the key outstanding issues. What we found is that the company was not prepared to make any substantial commitment in the areas of environmental protection and financial benefits."

A SEASON OF DISCONTENT

While both the Innu and the company tend to be tight-lipped about the IBA negotiations, a source close to the talks later revealed that, as a starting position, Inco had tabled an offer similar to an IBA that Falconbridge had signed 30 months earlier with the Québec Inuit. In return for access to the Raglan nickel-copper ore body in northern Ungava the company had agreed to pay the Inuit a 4.5 percent share of annual operating cash flow from the Raglan property. But, instead of offering the Innu the same percentage share, Inco was reported to have offered only the same dollar value. (The Raglan profit share is estimated to be worth $75 million to the Québec Inuit over the first fifteen years of the project.) But Voisey's is a much larger and, theoretically, at least, a much more lucrative undertaking, and the Innu quickly, and angrily, spurned what they regarded as a low-ball offer.

The Innu's minimal demand, their news release made clear, was at least 3 percent of net smelter royalties, which is the amount that Voisey's co-discoverers Verbiski and Chislett had been promised as "a discovery bonus." "A financial package in which the royalties payable to the Innu people are at least as high as the ones that the company will be paying to Archean (Verbiski and Chislett's company) is our absolute bottom line," Rich insisted. "The Innu people will never accept an IBA if they think that the two guys who stumbled on the discovery could receive more in royalties from the company than the people who actually own the land."

While Falconbridge evidently felt it could afford to share 4.5 percent of its annual cash flow from Raglan with the Québec Inuit, Inco faced the daunting prospect of matching whatever it paid to the Innu to the Inuit of Labrador as well, plus the discovery bonus. In total, some 9 percent of Voisey's profits would then have to be shared with parties other than company shareholders.

Rich went on to accuse the Voisey's Bay Nickel Company of treating the Innu "like children," and warned that "there is a real possibility that Innu people will return to the site," a thinly veiled threat to resume August's protest at the Anaktalak camp.

THE BOOTLEG EXPLODES

The rupture in the Innu IBA talks was yet another significant setback to the Voisey's Bay project, not least because Rich's angry language was indicative of the deteriorating relationship between Inco and her people, who would, after all, have to become something akin to real partners if the mine/mill development was ever to get off the ground.

But even as the Innu were issuing their news release in Davis Inlet on Monday, September 15th, Voisey's Bay was facing yet another potentially serious challenge on an entirely new front, in a courtroom in downtown Toronto. The St. John's environmental lobby was joining the fray, though its concern was not so much the mine/mill project in Northern Labrador as the planned smelter/refinery project in Argentia. With the financial backing of the Canadian Environmental Defence Fund, St. John's environmentalist Jim Brokenshire and his fellow members of the Citizens' Mining Council (CMC) of Newfoundland and Labrador were applying to the Federal Court of Canada to order a full environmental review of the $1.05 billion smelter/refinery scheme.

While the mine and mill development in Northern Labrador, expected to cost $350 million, was receiving a full-scale environmental review by a joint federal-provincial Environmental Assessment Panel, Brokenshire told the Toronto press, the much larger, more expensive, and potentially much more environmentally hazardous smelter-refinery complex was slated to receive much less public scrutiny. Known as a "comprehensive study," the assessment afforded no independent panel, no mandatory public hearings and no intervener funding, Brokenshire noted. Instead the billion-dollar project would receive only "a fast-tracked, behind-the-scenes, bureaucratic" review, the CMC president charged. "The environmental assessment was improperly split in two under a deal that excluded the 150,000 people of St. John's who live 60 kilometres downwind from the smelter site and who will be breathing Inco's air pollution," he added. "There is no point to building the smelter without the mine, and no point to building the mine without the smelter. Let's stop pretending they are separate projects and look at the whole picture so we can effectively protect the environment and human health."

Coincidentally, Inco was in court that Tuesday in Sudbury to face charges, which would eventually result in convictions, that it had violated the Ontario Water Resources Act by discharging contaminants from its Levack Mine tailings area into Grassy Creek, a tributary of the Onaping River, on February 23rd and March 2nd, 1994, and then by failing to inform the Ontario Ministry of the Environment about the spills.

A more enlightened company, benefitting from more sophisticated public relations advisors, might have gone a long way toward disarming its critics by welcoming the interest of the St. John's environmentalists, while maintaining that its smelter/refinery plans would withstand the most rigorous public scrutiny, but that it would, in any event, be happy to comply with whatever level of environmental review the courts, government, and public thought necessary. Instead Inco bristled, and, in the tradition of its two-fisted, not to say ham-handed, approach to regulation and the public interest, retorted with a typical, hardball response. David Allen, a Toronto-based company spokesperson, sniffed that under Canadian law a full environmental review was not required because the Argentia site, a former U.S. naval base, was already a "brown field" location, i.e., a heavily used industrial area and seaport. The clear implication appeared to be that, since Argentia was already heavily polluted by earlier industrial activity, there was little need for an exhaustive study of the impacts of one more industrial use. The company braced to fight the CMC's motion before the Federal Court along with lawyers from the Newfoundland government.

Newfoundland government lawyers were busy on another front that Tuesday as well. In St. John's a panel of three judges from the Court of Appeal of the Supreme Court of Newfoundland began to hear legal arguments on behalf of the Labrador Inuit Association and the Innu Nation as to why the infrastructure construction work should be permanently halted pending the completion of the report of the Environmental Assessment Panel on the full mine/mill proposal. As they would in the Federal Court case, lawyers from the Tobin government sided with lawyers from the Voisey's Bay Nickel Company in the proceedings, which would last for three days.

As if all that weren't enough, the *Globe & Mail's Report on Business* brought embattled Inco shareholders still more bad news that Tuesday morning: just below a story detailing the latest legal challenge to the Voisey's Bay project in the Federal Court was a story headlined "Indonesian drought hits Inco unit." The upshot of the piece was that the El Niño weather system in the Pacific Ocean had created the worst drought in Indonesia in 50 years. So severe had conditions become that water levels behind Inco's hydroelectric dams had dropped precipitously. The company's nickel smelter at Soroako, Indonesia, received 85 percent of its power from the dams, the article explained, and a shortage of hydroelectric power would in turn result in a curtailment of nickel production. Even the weather, it seemed, was conspiring to add to the nickel giant's sea of troubles.

Then, on the morning of Friday the 19th yet another bombshell dropped over the Voisey's Bay battlefield, which was becoming more littered and hotly contested by the day. The lead story in the *Toronto Star*'s business section that morning was a piece headlined "Voisey's Bay stalled." Star staffer Thomas Walkom reported that the Voisey's development "has been thrown into jeopardy following a decision by Inco Ltd. to reassess the project." While the project "probably won't be jettisoned entirely," Walkom noted, it "could be scaled back by as much as 50 percent." The company was also expected to announce "today that it cannot meet its December 1999 target of producing nickel concentrate" from Voisey's, Walkom revealed. "Inco has also decided to give up efforts to build an airstrip and road at the Voisey's Bay site, which it said were necessary to develop the project," the *Star*'s reporter added.

Could it be that Inco's senior executives were about to acknowledge, for the first time, the difficult realities that were besetting them, on the ground, in Newfoundland and Northern Labrador? Newsrooms across the country scrambled to "match" Walkom's "scoop," including my own at *Northern Life* in Sudbury. Reporter Keith Lacey succeeded in reaching Inco's chief financial officer Tony Munday at corporate headquarters in Toronto, and Munday hotly denied most of the salient points in Walkom's article. Suggestions that Inco was considering scaling back the Voisey's project were simply not true, Munday asserted, though he conceded that the company was "perhaps overly optimistic" in projecting a 1999 startup date for Voisey's ore. Munday also resorted to a time-honoured Inco tactic when dealing with media accounts it didn't like: discrediting the messenger. Walkom's account, Munday observed to Lacey, contained no attribution, but instead relied on unnamed "sources close to the decision." Inco had spent far too much money acquiring Voisey's Bay to simply walk away from it, Munday assured the Sudbury reporter.

Yet despite Munday's denials, most of the details of Walkom's exclusive would be confirmed by other company executives later the same day, creating huge headlines in the Bay Street press on Saturday morning. Even as the Inco chief financial officer was on the phone with Lacey, Inco CEO Mike Sopko was breaking the bad news about Voisey's Bay to an investors conference in London, England. The project would not begin production until the year 2000, Sopko confirmed, and the whole project was now indeed undergoing a reassessment. In Toronto Inco vice-president David Allen told the *Financial Post* that the review was likely to include the 270-million-tonne annual production level for Voisey's Bay. "Although details were sketchy, most observers assume Inco is planning lower nickel production from the target," the *Post* concluded. Allen told the *Globe* that Inco had decided to scrap plans for the temporary road and airstrip. "We thought we stretched the timeline as tight as we could," Allen conceded in a statement that could come as news only to anyone unfamiliar with the climate of Northern Labrador.

At the same time, the company issued a faxed press release that included a carefully worded nuance that most of the financial press unaccountably omitted from their stories the next morning. The company "now believes that initial production from the Voisey's Bay mine and mill facilities will be delayed by *at least* one year or until late 2000 *at the earliest,*" the news release noted (emphasis added). Not only were company officials at last squaring their public utterances with reality on the ground in Labrador and Newfoundland, they were also leaving themselves open to the possibility that even the latest deadline might not hold, as indeed it would not.

Bay Street, which had applauded Inco's acquisition of the Voisey's Bay property just a year earlier, wasted little time in passing judgment on Inco's predicament, and within a few hours it was clear that the sharks were in the water. Inco shares dropped $2.00 before the closing bell that Friday, to a new 52-week low of $33.60. "For every year the project is postponed one analyst estimates that it knocks $1.90 off the [Inco share] price," the *Globe's Report on Business* noted helpfully in a front-page, above-the-fold box labeled "Investor Impact," lest any Inco shareholder within the paper's reach might miss it.

Sopko and Allen blamed "the environmental review process and problems in reaching agreements with the Aboriginal communities" for the delay, the *Globe* reported. "We didn't anticipate at all that we would be subjected to a new form of environmental review process put in place by the two governments and the two aboriginal peoples," intoned VBNC executive vice-president Rick Gill, rather disingenuously. "It's just too much. This is unprecedented."

For his part, the *Globe's* ultra-conservative corporate apologist Terence Corcoran blamed "a long line of freeloaders and panhandlers" for "stalling progress and threatening to put the whole project in jeopardy if they don't get a handout." The line, Corcoran made it clear, included the people of Newfoundland, who wanted to tax the project unfairly, the Citizens' Mining Council, for their insistence on an environmental review, and the Innu and Inuit, who were guilty of just plain greed. Corcoran's "freeloaders and panhandlers" did not include Robert Friedland, who was worth billions in cash and stock options due to the sale of Voisey's to Inco, or Falconbridge, which had pocketed a cool $100 million just for entering the bid that drove the price up, or prospectors Verbiski and Chislett who, Corcoran quoted Katie Rich as saying, stood to earn $490 million over 20 years on their discovery bonus. "The objectives of the native groups seem pretty clear," Corcoran noted darkly, implying that they were guilty of unreasonable avarice for seeking a share of the proceeds about to be derived from land that, after all, belonged to them.

And then, on Monday, September 22nd, came the unkindest cut of all. In a unanimous judgment the three justices of the Newfoundland court ruled that the Newfoundland minister of environment and labour had indeed exceeded his authority in issuing work permits to Inco for its "Exploration Support Work" program. The justices quashed the permits, effective immediately, thereby effectively dashing any hope the company might have had of completing any further work on the site until the Environmental Assessment Panel had completed its work.

The 33-page judgment was an articulate, at times even eloquent, defense of the environment, and the environmental assessment process, and it is worth citing at some length. It opened with a quote from an earlier Supreme Court of Canada judgment in *Friends of the Oldman River v. Canada* that described protection of the environment as "one of the major challenges of our time." In this statement, the Supreme Court of Canada encapsulates the critical need of reconciling the use of the Earth's natural resources with the protection of the environment," Justices W.W. Marshall, G.L. Steele and J.D. Green wrote.

"The need to rationalize these imperatives is a phenomenon of relatively recent origin. This is because for most of the history of humankind the development and sustenance of life has been molded and controlled by the environment. As Rachel Carson has pointed out, it has only been in the last century that the relationship has been reversed to the extent that humans now possess the power to mold and change the environment in significant ways (*Silent Spring*, Crest Paperback edition, 1962, p. 16). The web of life, which contains and controls the interdependence of living things and beings, both with respect to each other and to their physical surroundings, is not static. Change in one area may profoundly affect life and habitat in other areas and may even threaten its existence in ways that cannot be immediately foreseen.

". . . In this Province, as elsewhere, society has been left to grapple with the deleterious, and at times tragic, effects of unbridled development on the health and security of its residents and upon the environment. The recent experience of the devastation of the fishery through over-exploitation bears stark witness to the consequences of the impact which the pace of humankind's activities, especially those driven by economic forces, can have."

Still, the Court continued, there was a need to balance environmental protection against "the economic and social benefits that flow from the production of these resources. Legitimate concerns of meaningful employment and security for families are at stake. This is a reality that must also be taken into account along with environmental considerations."

While it is imperative that development and investment must also be weighed in the balance, the managers of investment "cannot be allowed to control the agenda without regard to competing environmental interests.

"No natural resource is a forbidden fruit. . . . The challenge is to temper the refrain advocated by developers from time to time 'to develop or perish', by assuring that it does not re-echo amongst future generations as 'develop and perish.'"

One of the primary means by which governments have attempted to balance economic development with environmental protection is the statutory requirements for environmental assessment, the Court noted. "It appears just plain common sense to require development of resources to await the relatively short time that will be taken to allow adverse environmental effects to be assessed and mitigated, if not eliminated."

The Court then went on to review the specifics of the case and ruled that the trial judge had erred in allowing the "temporary" infrastructure work to be exempted from the environmental review process under the MOU, and that Inco's semantic sleight-of-hand in labeling the work "temporary," as opposed to "permanent," should not allow the construction to escape the assessment regime. "It appears transparently artificial to maintain a road and an airstrip to be used for one purpose which lies on the same land when subsequently used for another, albeit improved [purpose]."

Newfoundland Minister of Environment and Labour Kevin Aylward, the Court concluded, had exceeded his authority in granting the work permits, "and has failed to comply with the MOU." While it recognized that investors in Voisey's Bay "are also furthering pressing social and economic concerns of substantial numbers of Newfoundlanders and Labradorians, and their families, who have legitimate expectations of badly needed meaningful employment, the Inuit and Innu may also be viewed as representing general vital interests."

The Court then awarded costs to the Inuit and Innu and declared "the Minister's decisions of May 22, 1997 and July 2, 1997 are hereby quashed and declared to be of no force and effect."

It is probably no exaggeration to say that the Court's decision was of historic importance and appeared destined to establish an important precedent in all future applications of the Canadian Environmental Assessment Act. It is also worth noting that the Crown elected not to appeal the landmark ruling. But for Inco the decision was little short of catastrophic. Even as the Bay Street sharks circled hungrily in the water around the wounded nickel giant, demanding concrete results, some *progress* at Voisey's Bay, Inco insiders understood at once the implications of the ruling: not only was the remainder of 1997's construction season now a loss, but the next season would almost certainly be lost, too, because the Environmental Assessment Panel was not expected to complete its work before the fall of 1998. Given a three-year construction period from the start of infrastructure work to the completion of the mine/mill complex, production could now not be expected before 2001, at the earliest, and even this date was predicated on swift resolution of comprehensive land claims and IBA talks, not to mention a host of unresolved issues relating to the smelter/refinery complex in Southern Newfoundland.

In hindsight, the predictions of Christine Cleghorn, Voisey's Bay Assessment Coordinator for the Innu Nation, had proven far more accurate than those of Mike Sopko and Stewart Gendron. Although she sat in a modest office in a remote community far from the money and power of Bay Street, Cleghorn's words in April now appeared prophetic. Her dismissal of Sopko and Gendron's timelines, her promise that the Innu Nation would, one way or another, prevent the infrastructure work, had all come true. The Inco Environmental Impact Statement that Gendron had promised, in April, would be delivered by July, and was still months away from actual completion.

If it had refrained from pushing ahead with the infrastructure work and raising the expectations of Bay Street, and if it had respected the wishes of its Aboriginal neighbours, Inco might, in theory at least, have been able to complete its work unimpeded in the summer of 1998 in the event of successful talks with the Innu and Inuit. Now, even that unlikely possibility was formally foreclosed, by Court order. By the end of the season of discontent the company had lost all control over the timing and pacing of its Voisey's Bay development. Inco was now at the mercy of myriad factors over which even it, still the largest nickel producer in the world, with a global reach and seemingly infinite resources, would have less and less command as the months wore on. The standoff at Voisey's Bay had now begun in earnest.

Monique Rich at Anaktalak Bay.

(left to right) Larry Innes, Chief Prote Poker (facing), Katie Rich (back turned) and Chief Paul Rich Jr. plan strategy at the upper protest camp.

Putting it all in perspective—an editorial cartoon by Innu protester Andrew Penashue.

Non-Native protester sent in effigy by non-native supporters unable to attend protest.

Innu protesters marching on the Inco camp.

Innu youth salute the RCMP at Anaktalak Bay Camp.

Innu youth climb atop trailers at Inco camp, frustrating RCMP officers.

LEFT: Innu marchers on the newly constructed road at Anaktalak Bay camp. Roadwork was suspended by the protest and terminated by the Newfoundland court of appeals.

BOTTOM: Aerial view of mining camp at Anaktalak Bay.

Voisey's Bay at Last: Although Inco refused to help, the author makes it to the company's campsite at Anaktalak Bay on October 23rd, 1996.

Place of the Boss: Although it looks picturesque enough, Davis Inlet-Utshimassit is universally depised by its 700 Innu residents because of its isolation and because of the shortage of fresh water.

LEFT *Iron Fist and Velvet Glove:* Like generations of Sudbury reporters before him, young Peter Evans of Nain's OkalaKatiget Society ran afoul of Inco for publishing stories the company didn't like.

BOTTOM *Granny Flats, Davis Inlet Style:* What the author first mistook as a shed building boom in Utshimassit turned out to be an attempt to alleviate a critical housing shortage in the Innu community.

Disarmed and Dangerous: Katie Rich, President of the Innu Nation, is as at home on her snow machine as she is conducting the in-your-face negotiations with corporate and government leaders for which she has become legendary.

Voisey's as Vortex

Fall – Winter 1997 – 98

11

Nickel & New Caledonia: They Shoot Kanaks, Don't They?

"We operate all over the world and we've never experienced this kind of difficulty before."
  – Inco vice-president David Allen, Canadian Press dispatch,
    September 19, 1997

NO REMARK, with the exception of some of the bawdier cards at my 50th birthday party, occasioned quite so much merriment among Sudburians of my acquaintance as the statement from David Allen. It was greeted with hearty roars of laughter by several people, and one friend, a long-time Inco watcher, called me back, long distance, to have the exact wording read to him over the phone.

The "difficulty" that provoked the remark was, of course, the necessity to deal with environmental regulations and Aboriginal land claims in Voisey's Bay. To abide by the rule of law, in short. The

VOISEY'S AS VORTEX

unintended humour in Allen's remark is the fact that, as most every Sudburian knows, Inco's last two offshore forays, into Guatemala and Indonesia, put the company into bed with two of the most repressive political regimes on the planet. Aboriginal land claims? Environmentalists? Union organizers? Land reformers? Shoot the bastards! Oh, not the company itself, of course, but the ruthless thugs who control the two countries outside Canada where Inco has established, or tried to establish, integrated mining/milling/smelting operations.

Think I'm exaggerating? In just one of its more notable atrocities the government of Indonesia is widely reported to have shot, slaughtered or killed by starvation 100,000 women, children and men in East Timor when it invaded that tiny country in December 1975. That's out of a total population of 600,000. The invasion occurred just before Inco elected to invest heavily in its P. T. Indonesia mining and smelting operation.

In Guatemala, thousands of citizens, many of them Maya Indians, were killed in a bloody civil war that raged in that country while Inco was establishing Exmibal, its ill-fated subsidiary. In the Lake Izabal-Zacapa region alone, where Exmibal was located, 3000 Guatemalan peasants were killed in a "pacification campaign" led by Colonel Carlos Arana, who was later nicknamed "The Butcher of Zacapa" for this campaign. His picture appeared in Inco's 1973 annual report, by which time he had become president of Guatemala. Arana's sanguinary efforts are widely regarded as paving the way for Inco's Guatemalan investment. After all that, and some $200 million in cash derived largely from its Canadian operations, Inco mothballed Exmibal in the early 80's. Despite the expenditure of so much blood and treasure, Inco's Guatemalan operation hasn't produced so much as a thimbleful of nickel since.

Many historians trace the roots of the Guatemalan conflict to the CIA-engineered overthrow of the democratically elected government of Guatemalan president Jacobo Arbenz in 1954. John Foster Dulles was U.S. Secretary of State at the time, and his brother Allen was head of the CIA. Both men, of course, were former directors of Inco... but I digress.

You'd have to go some way to find an investment climate that has a worse human rights record than either Guatemala or Indonesia, but in announcing that it planned to spend $50 million in New Caledonia while reconsidering the whole proposition at Voisey's Bay, Inco managed to find just such a place. New Caledonia is a cluster of islands in the South Pacific roughly 1500 kilometres off the east coast of Australia. The largest of the islands, Grand Terre, is sizeable—400 kilometres long and 48 kilometres wide—making it the largest land mass in Micronesia. It is also, not coincidentally, said to be almost solid nickel.

Actually, New Caledonia and the Canadian nickel industry share a long history. In 1901, even before Inco was born, the Canadian Copper Company, based in Copper Cliff just west of Sudbury, came under considerable pressure from some nationalist-minded Canadian politicians and editorialists to refine its Sudbury ores in Canada. Canadian Copper's entire smelter output was then being shipped in matte to New Jersey for refining. "Oh no no no," the company's American owners said, "we couldn't possibly refine Canadian nickel in Canada and make a profit. And if you try to make us, we'll just mine our nickel in New Caledonia."

Nickel mining in New Caledonia predates Sudbury's industry. The mineral was first discovered there in 1864, and a French concern, Société Le Nickel (SLN), began large-scale mining operations in 1876. For a good part of this century Sudbury and New Caledonia ranked one and two in world nickel production, until the Soviet Union entered the picture in a big way after the Second World War.

The populations of the two communities are almost identical, too. There are about 170,000 people living in New Caledonia today, roughly the same number as in Sudbury. Yet despite the shared history and industry, only one Sudburian of my acquaintance has ever visited New Caledonia. The year was 1970 and

VOISEY'S AS VORTEX

Homer Séguin was the president of United Steelworkers of America Local 6500 at Inco.

The employees of SLN had let it be known that they wanted to form a union, and Séguin and USWA staff rep Emil Vallée had been dispatched to New Caledonia by the Steelworkers to assist the workers there. "The first thing we noticed as our flight approached [New Caledonian capital and SLN base] Nouméa [pronounced Noo–MAYAA] was the terrible environmental damage the industry had done," Séguin recalls.

"You could see the smelter effluents turning the ocean black far out to sea, there were only chimneys for smelter stacks, and the strip mines had created tremendous erosion from the tropical rains. Instead of building roads that might have helped develop the country the company had built these aerial tramways, big buckets moving along cables suspended from huge towers, to bring the ore into the smelter.

"The hills around Nouméa were all bare and brown from the smelter fumes and erosion, and we could see what looked like police or soldiers, on the hillsides, even from the air."

Once on the tarmac Séguin and Vallée, and Vallée's wife, who was six months' pregnant, noticed that hundreds of New Caledonians, SLN workers presumably, had turned out to greet them. But between the airplane and the crowd stood hundreds more French gendarmes or soldiers, packing submachine guns.

"They're waiting for us, Emil," Séguin remembers exclaiming to his colleague, disbelief still audible in his voice almost 30 years later.

"Naah," replied the Steel staff rep. "It's just tight security, is all."

The Canadian trio got no further than the airport terminal, where they were summarily refused entry into the country before being physically dragged out of the airport, across the tarmac, and back onto the plane that had just flown them in. They spent the next 30 days in New Zealand waiting for diplomatic pressure on France, which rules New Caledonia with an iron hand, to open their way to Noumea. It never happened.

"That's just so French," snorted Donna Winslow when I repeated Séguin's story to her. A professor of sociology at the University of Ottawa, Winslow is a Canadian expert on New Caledonia. New Caledonia, Winslow explained, has a history of Aboriginal resistance, and brutal repression, that makes the armed standoff at Oka look peaceful in comparison.

To its original inhabitants, the five islands of the New Caledonian archipelago are known collectively as Kanaky, meaning homeland of the Kanak people. The Kanak had been living on the islands for 30,000 years and numbered some 70,000 when they were "discovered" by Captain James Cook in September 1775. Oh happy day. Kanaky was annexed by the French in 1853 and then promptly turned into a penal colony. An estimated 20,000 French convicts were transported to Kanaky between 1864 and 1897. Nickel was discovered there in 1864, and the Kanaks were relegated to "indigenous reservations" to facilitate the opening of the mines.

By 1921 the Kanak had been dispossessed of all but 10 percent of their traditional lands. Armed rebellion, French military repression, and disease had reduced their population to approximately 27,000 people. Today Kanaky is the world's "last bastion of colonialism," according to Winslow. "It's France's *apartheid* government in the South Pacific."

And when Winslow uses the word "colonialism," she's not kidding. It wasn't until 1951 that the residents of Kanaky were finally given the right to vote by Paris. They elected Maurice Lenormand of the multi-racial Union Caledonienne to represent them, but he was thrown in jail for a year following a June 1958 armed uprising by French settlers, or *colons*.

Since then France has done everything in its power to prevent Kanaky independence, up to and including political assassination and murder. On September 19, 1981, Pierre Declercq, secretary-general of the Union Caledonienne, was shot and killed in his home. His death triggered a renewed independence movement, led by a former Catholic priest, Jean-Marie Tjibaou. Tjibaou was heavily influenced by an earlier generation of Paris-educated national liberation leaders and thinkers like Ho Chi Minh and Frantz Fanon, and he played a key role in organizing the Front de Liberation Nationale Kanak Socialiste or FLNKS in 1984.

Winslow wrote her doctoral thesis on the Kanaky independence movement, did postdoctoral research on the impact of mining in Kanaky, and served as Tjibaou's executive assistant for ten years. For a Canadian, hers was a unique vantage point from which to observe the tragedy that was about to unfold.

"The FLNKS was a very unique organization comprised of five political parties, the trade union and the women's organization," Winslow explained. On December 1st, 1984, the FLNKS proclaimed a provisional government and organized roadblocks. French President François Mitterrand dispatched a personal envoy to Kanaky to devise a plan for self-government two days later, but on December 5th, French settlers armed with automatic weapons ambushed a group of 17 unarmed FLNKS. Ten Kanaks were killed, including two of Tjibaou's brothers. Two years later a jury of all whites except for one Indonesian acquitted the *colons* responsible.

On January 12, 1985, police shot and killed Eloi Machoro, minister of the interior in the provisional government, along with his aide. At the same time the Mitterrand government announced sweeping reforms that would grant Kanaky a measure of self-rule. Those reforms were summarily rescinded, however, with the election of the conservative government of Jacques Chirac, whose government proclaimed the Kanak people non-existent and announced that only French citizens lived in New Caledonia.

In April 1988, a group of 40 Kanak youth from the island of Ouvéau attempted to hoist the Kanak flag over the local police headquarters. A firefight ensued, four French police were killed, and 27 gendarmes were taken hostage in a grotto near the village of Gossanah. At dawn on May 4, 1988, 300 French troops surrounded the grotto and poured some 10,000 rounds into the place. Thirteen of the young Kanak men were killed outright and another six were executed following their surrender.

The Ouvéau Massacre, as it is now known, led to the August 1988 Matignon Accords, in which France pledged that a referendum on Kanaky independence would be held in ten years' time. Jean-Marie Tjibaou negotiated the Accords on behalf on the FLNKS.

The Kanak mourning period is one year, Winslow explained, and on May 4, 1989, while attending a commemorative service in Ouvéau, Tjibaou and his deputy Yeiwene Yeiwene were gunned down by Djoubelly Wea, an erstwhile member of the FLNKS. Wea, in turn, was immediately shot and killed by Tjibaou's bodyguards. This, then, is the climate of historical and political instability that Inco, and the Sudbury employees who will doubtless be dispatched there to help open its new mine, will enter. A sovereignty-association agreement is still scheduled for late 1998, according to Winslow, though the FLNKS is in disarray following the assassination of Jean-Marie Tjibaou and the pro-independence forces are divided on the wording of the question. Although they comprise 52 percent of Kanaky's population, half of the Kanak people are under the age of 20, Winslow adds, and are therefore ineligible to vote. Oh, and just in case you were wondering, the Jacques Chirac whose government attempted to extinguish Kanaky's long-standing aspirations for self-government is the same Jacques Chirac, now president of France, who has assured Québec Premier Lucien Bouchard that France would recognize an independent Québec.

12

Sui Generis (1)

It wasn't until I met Steve O'Neill in the fall of 1997 that I fully understood the magnitude of the Aboriginal land issues confronting Inco in Voisey's Bay, or indeed any other Canadian mining company hoping to exploit resources in Canada's far north in the years to come.

O'Neill is the Sudbury lawyer who cautioned Inco CEO Mike Sopko about Aboriginal claims to Voisey's Bay at a conference back in November 1995, and he knows whereof he speaks. A partner in the law firm of Miller Maki, O'Neill is one of a handful of Ontario lawyers who has turned the representation of First Nations around land issues into a full-time job.

I was astonished to learn the following two facts—facts that Steve O'Neill knows very well:

- By the year 2000 fully one-third of Canada's entire land mass will be under Aboriginal ownership or control.
- An estimated 60 percent of this country's mineral potential will lie under that one-third of Canada.

It would be tempting to declare that this is an asset shift of immense and historic proportions, but as O'Neill explained to me over lunch in a Chinese restaurant, Canadian courts are increasingly of the opinion that, in recognizing Aboriginal land claims, they are merely acknowledging a status quo that has existed for a long, long time.

"The interest that Aboriginal people in Canada have is *sui generis*," O'Neill begins, "that is, really unique." Aboriginal rights were, moreover, acknowledged and entrenched in Section 35 (1) of the Canadian Constitution, which became law in 1982. Not surprisingly, O'Neill knows those few, simple, and utterly clear words by heart: "The existing Aboriginal and treaty rights of the Aboriginal peoples of Canada are hereby affirmed and guaranteed."

What this means in terms of land ownership is a still-emerging legal minefield into which Inco has not so much tiptoed as stomped. Crucial to the question is the undisputed fact that in much of Canada (including Labrador) no treaties were ever signed between First Nations and the Crown. In a series of decisions, most notably the 1984 case of *Guerin v. the Queen*, the Supreme Court of Canada has ruled that, in the absence of a voluntary surrender by treaty or legislation, Aboriginal people may possess certain rights to Crown land. "The Crown does have underlying or radical title to Canadian lands," O'Neill explains, "but Aboriginal land rights burden that title."

Who owns this or that piece of Crown land or the minerals under it can be resolved in the courts, of course, but few of the parties in a given dispute want to go there, O'Neill observes, because the process can take years, or even decades, because it's expensive, and because the outcome is far from certain. Writing in the *Canadian Bar Review*, Osgoode Hall law professor Brian Slattery summed up the post-Guerin landscape this way: "Aboriginal title is a legal right that can be extinguished only by native consent or by legislation. The effect is to shift the burden of proof to federal and provincial governments.

"They must now show that Aboriginal land rights were lawfully extinguished in the past or acknowledge their continuing existence.

SUI GENERIS (1)

Where the rights were wiped out by legislation, the decision implies that compensation should have been paid."

Does this mean what I think it means? I asked O'Neill. That Inco may have paid $4.3 billion for a piece of Crown land to which the vendor (Robert Friedland of Diamond Fields Resources) may not have had clear title in the first place?

O'Neill chewed his sweet and sour thoughtfully before answering. "The point can be made that their (Inuit and Innu) consent will be required for this development to go full steam."

How in the world could this have happened? I queried. If Mick Lowe can learn this over a plate of chicken chow mein at Sudbury's Lodo Restaurant, how could Inco, which retains one of Bay Street's most prestigious law firms, have failed to read this not-so-fine print?

O'Neill shrugged. This does not mean, he hastened to add, that Voisey's Bay will never be developed. But it may mean that the Innu and Inuit have just as good a claim on Voisey's under Canadian law as Inco, $4.3 billion notwithstanding.

What it almost certainly means is that Inco, and other mining giants wanting to do business in the Canadian north, will have to come to terms with the land's Aboriginal inhabitants before mining can begin. It *can* be done, and O'Neill's business is doing precisely that; hammering out agreements, on behalf of First Nations, with governments and private companies. He has successfully concluded a half-dozen agreements in northeastern Ontario, including a deal that saw a payment of $7.5 million and the return of 24,000 hectares of land to the Garden River reserve just east of Sault Ste. Marie.

O'Neill also negotiated Inco's first-ever agreement with a Canadian Aboriginal group, an August 1996 undertaking with the Wahnipitae First Nation concerning the decommissioning of the Whistle open pit mine north of Sudbury.

It can be done, but O'Neill cautioned that a deal as large and complex as Voisey's Bay could very well delay—big time—Inco's already much delayed timetable for nickel production in Labrador. The Garden River claim, which was settled in 1994, was first lodged in 1972. The Labrador Inuit Association began negotiating its land claim with the Newfoundland and federal governments more than 20 years ago.

"Maybe there's a lesson for the governments here," O'Neill concluded with an ironic smile. "Settle these claims *before* the nickel is discovered."

■

O'Neill's admonition was not lost on a small group of Torontonians who may well bulk large in the battle over Voisey's Bay. Although urban-centered and non-Aboriginal, these individuals have founded two organizations, the Friends of the Lubicon and the Voisey's Bay Innu Rights Coalition (VB/IRC), aimed at supporting the land claim struggles of Aboriginal peoples thousands of kilometres away.

Anyone with a stake in Voisey's Bay or Inco also had an interest, whether they knew it or not, in a legal battle that pitted these same activists against a huge Japanese forestry conglomerate. I first learned of the case at about the same time as my lunch with Steve O'Neill, after being invited to Toronto by members of VB/IRC to lead a teach-in on the history of Inco. The court case, known formally as *Daishowa v. Friends of the Lubicon*, was being argued in the fall of 1997 in the Ontario Court, General Division, in Toronto.

Daishowa Inc. is a family-owned Japanese forestry multinational (1996 profits well over $1 billion) whose size makes even Inco look picayune. In 1989 the Alberta government granted the company a license to harvest timber on lands in northwestern Alberta that had been claimed by the Lubicon Crees.

The Lubicon Crees, who have been attempting to negotiate a formal land claim settlement with the federal and Alberta governments since the 1930's, were vaulted into national prominence back in the 1980's when oil and natural gas companies invaded land

SUI GENERIS (I)

the Lubicon claimed was theirs. Exploration activity had a disastrous impact on the Lubicon people, driving away the moose herd and increasing welfare dependency and disease, while eventually producing $8 billion worth of petroleum for the companies in question.

This apparent injustice caused a group of Toronto social activists to take up the Lubicon cause and form a protest group called Friends of the Lubicon, or FoL for short. FoL decided to target Daishowa's activities in urban Canada as a gesture of solidarity, and to apply pressure on industry and government to settle the Lubicon land question once and for all. The Toronto group approached some of Daishowa's customers and urged them to purchase their brown paper bags from alternate suppliers unless and until the company agreed to hold off its logging activities in northwestern Alberta, pending the resolution of the Lubicon land claim.

This David versus Goliath action produced a surprising result: the company did, in fact defer logging on Lubicon lands, on a year-to-year basis. It also accused the FoL of costing it $11 million in lost sales, and sued the activists for millions of dollars in damages. Despite its clear importance to the Charter right to free speech, the court case received scant media attention, so on Monday, September 15, 1997, a group of prominent Canadians, including David Suzuki and former CBC boss Patrick Watson, called a news conference in Toronto to focus attention on the case and to condemn Daishowa.

"Here we have a company based in Japan, seeking to exploit a Canadian resource, receiving Canadian subsidies, using Canadian courts to persecute Canadians for exercising their right of free speech," summarized Suzuki at the news conference, which also received shamefully little press coverage.

What did any of this have to do with Inco and Voisey's Bay? The FoL activists, many of them, were also prime movers in the Voisey's Bay/Innu Rights Coalition. I soon discovered that they were a youthful, energetic group at pains to make sure what happened to the Lubicon Cree in Alberta doesn't happen to the Innu Nation in Northern Labrador.

VB/IRC made its first public appearance at the April 1997 Inco annual meeting in Toronto. They leafleted the meeting to heighten shareholder/investor awareness about Innu and Inuit claims to Inco's Voisey's Bay property. Several ethical investment funds that hold Inco shares contacted the group after the meeting to learn more about the company's activities in Northern Labrador, opening a dialogue that still continues.

So far, no one has asked the funds to dump their Inco shares, FoL and VB/IRC activist Kevin Thomas told me, for the simple reason that the Innu Nation has not chosen to escalate its battle around Voisey's Bay in this way—at least not yet. But these Toronto activists, mostly in their 20's, are very savvy in the ways of environmental science, the media, research, e-mail, the Internet, and the urban environment, and they represent a potentially troublesome fifth column on Inco's own home base—Bay Street.

They're also keenly aware of the amazing success enjoyed a couple of years ago by the James Bay Cree in stopping further expansion of Hydro-Québec's hydroelectric projects in Northern Québec. Cree leader Matthew Coon Come took his people's struggle against the project to Wall Street and to the huge American utilities Hydro-Québec was banking on to buy its power. The Americans pulled the plug after listening to Coon Come's pleas, and all bets—and the planned expansion—were off.

If they chose to, could two Aboriginal nations in Northern Labrador scupper the entire Voisey's Bay development by contesting the matter in the urban/financial arena, a tactic used to such good effect by the Friends of the Lubicon and the James Bay Cree? If land claims and impact and benefits talks involving the Inuit and Innu, Inco, and the federal and provincial governments falter, a few thousand Aboriginal people, combined with their urban allies, might just render the Voisey's Bay project too expensive, in terms of share price, timelines, and public, global, image for Inco to swallow.

Part of the likelihood for such an eventuality was being tested in the case of *Daishowa v. Friends of the Lubicon*. If companies like Inco, Daishowa, and Hydro-Québec are the Goliaths in these contests to exploit remote rural resources on Aboriginal lands, then the Innu, Inuit, Lubicon and James Bay Crees, and their urban allies, are certainly the Davids. But these latter-day Davids, like the legendary Biblical king, have shown they can fell even corporate Goliaths with their unassuming slingshots. Or would the Canadian courts strip David of even his slingshot?

In early April 1998, the Court ruled against Daishowa and in favour of the FoL's right to protest. Initially the company vowed to appeal the decision to the Supreme Court of Canada, and the Friends of the Lubicon resumed their boycott of Daishowa products. But by June the company had second thoughts and agreed to suspend logging on Lubicon-claimed lands indefinitely, pending resolution of Aboriginal claims in the region. Chalk one up for the slingshot.

∎

The importance of treaties to resolve land tenure questions and facilitate development was brought forcefully home to Sudburians on Monday, December 1st, 1997, when our community entered a new era in nickel production. That was when Falconbridge Ltd. produced the first pound of nickel concentrate from its new mine/mill operation at Raglan. Located at the northern tip of the Ungava Peninsula in Northern Québec, Raglan will play an important part in Sudbury's long-term future as a metallurgical centre, guaranteeing our role, at Falconbridge at least, well into the 21st century.

In contrast to Inco's uncertain land tenure in Northern Labrador, Falco's Raglan development had been enabled by a comprehensive land claims agreement—one of the few so far concluded in Canada in the modern era—the James Bay and Northern Québec Agreement (the JBNQA) of 1975.

The Raglan ore body will boost Falco's annual nickel output by close to 50 percent, and all of the concentrate milled at Raglan will be sent to Sudbury for smelting before it is shipped on to the company's refinery in Norway, where it will be refined into finished nickel. The company plans to invest $158 million over five years to upgrade its smelter in the town of Falconbridge, in large part to accommodate Raglan concentrate. While that sizeable capital investment will not create additional jobs, it will at least ensure long-term employment at the smelter for decades to come.

In a very real sense, too, the beginning of production at Raglan meant that Sudburians had new partners in production—the Nunavik Inuit people of Northern Québec—the same group whose elders I had encountered in the restaurant in Nain 15 months earlier. Half-a-dozen Falco employees from Nunavik had travelled to Sudbury for training at the company's operations here before the mine opened, and more will surely follow.

In February 1995, the Makivik Corporation, the business arm of the 8500 Nunavik people, signed a landmark Impact and Benefits Agreement (IBA) with Falconbridge, which opened the way for the company to spend $500 million developing the Raglan mines and mill. The IBA, which numbers 233 pages including appendices and annexes, contains terms and provisions for the Québec Inuit far beyond anything that Sudburians have ever been able to extract from either of our locally based nickel giants.

Among other things the IBA guarantees:

- a corporate commitment to Inuit employment and training;
- a joint Raglan Committee, composed of three Makivik representatives and three from the company, which will have sweeping powers to resolve on-site disagreements, especially regarding the environment;
- a Falconbridge commitment to refer disputes that cannot be resolved by the Raglan Committee to binding arbitration;
- a layoff provision that protects Inuit employees regardless of

seniority, provided requisite technical skills are available;
- an inside-track purchasing policy for Inuit-owned businesses to sell services and material for use on the Raglan site;
- a Makivik seat on the board of Directors of Société Minière Raglan du Québec Ltée, the wholly owned Falconbridge subsidiary that will run the Raglan operation;
- and a 4.5 percent cut of the profits from the Raglan operation.

The Sudbury smelter is mentioned at some length in the IBA, and it makes Sudburians genuine partners with the people of Nunavik. Future costs of smelter maintenance and improvements will be prorated, and Raglan will pay a share, based on its percentage of annual smelter throughput. At the same time, wage or benefit increases that make their way to the bottom line will presumably reduce profits for Makivik and the Nunavik people.

The Raglan IBA was all the more remarkable because Makivik negotiators shrewdly leveraged the most tenuous land tenure position—a category three claim concerning environmental impacts on hunting and fishing plus certain offshore rights under the JBNQA—into an agreement that far surpassed five decades of collective bargaining agreements between Sudbury's nickel companies and three of Canada's most powerful trade unions, the United Steelworkers, the Canadian Autoworkers, and the old Mine Mill union.

It would be easy—but wrong—for Sudburians to be envious and complain about this IBA, which affects Sudbury's future but into which they had no input whatsoever. Good on Makivik and Falconbridge both for negotiating this agreement, which ushers in a whole new era of corporate-community cooperation. The correct approach should be to learn from this agreement and ask how it might serve as a model for the future of Sudbury and other Canadian mining communities.

The Raglan IBA is widely regarded as a model that Inco will be expected to meet, if not exceed, not once, but twice—with both the Labrador Inuit and Innu—before it can begin serious development at Voisey's Bay. If that happens, both Inco and Falconbridge will have acceded to an unprecedented level of community, in this case Aboriginal, input and control of their ongoing day-to-day nickel operations.

The big difference, of course, is that the Nunavik, Labrador Inuit and Innu peoples have some kind of legal claim on the land under which desirable ore bodies lie, giving them considerable leverage with both companies. But Falconbridge vice-president Thomas Pugsley told me that the company also felt a moral obligation to the people of Nunavik, especially in view of the Whitehorse Mining Initiative. Signed in September 1994 by some of Canada's leading mining executives, environmentalists, trade unionists, mines ministers and Aboriginal leaders, the WMI commits the industry "to expand the opportunity for meaningful and responsible participation . . . by local communities in decision-making processes that affect the public interest."

Both Inco and Falconbridge are members of the Mining Association, and Falconbridge, in particular, played a key role in negotiating the Whitehorse Mining Initiative. The original document was signed by, among others, Alex Balogh, president of Noranda, the parent company of Falconbridge, and by Rick Briggs, former president of Mine Mill Local 598 at Falconbridge in Sudbury, among others.

If the industry's enlightened new policies of consultation, shared decision-making, and community profit-sharing are good for the Inuit and Innu and Nunavik peoples, why are they not also essential for the people of Northern Ontario, who have generated so much wealth for Inco and Falconbridge and whose labours have underwritten costly expansion elsewhere?

■

In late October of 1997 it appeared that there was good news on yet another land claim front, this one having a direct bearing on Voisey's Bay. The Labrador Inuit Association had, at long last, reached what its leaders considered to be an adequate, if rough, blueprint for a comprehensive land claims settlement for the traditional Inuit lands of Northern Labrador. The resulting document was not an Agreement-in-Principle, or AIP, but rather an agreement to Agree-in-Principle, the LIA leaders explained.

The breakthrough was not wholly unexpected; the LIA had been negotiating its land claims for 20 years, after all, and many observers had long anticipated that the Inuit would settle before the Innu, thus paving the way for an Inuit Impact and Benefit Agreement with Inco, which would, in turn, remove one of the key logs in the logjam that the Voisey's Bay development had become by the fall of 1997.

There were a number of reasons why the Inuit had been expected to settle first. While the Inuit had been negotiating for two decades, the Innu had only begun their land claims negotiations in 1990. The Inuit claim, which had been treated in a desultory fashion by Ottawa and St. John's, had, of course, received renewed impetus from the mineral discovery at Voisey's Bay. Suddenly it was in the government's interest to settle in order to facilitate development, and the long-outstanding Aboriginal claims to Labrador were "fast-tracked" and moved nearer the top of the bureaucratic agenda.

There was also an expectation, in both Newfoundland and Labrador, that the Inuit would prove more pragmatic, and perhaps more tractable, than the Innu. While the latter had been exposed to Euro-Canadian influence only relatively recently, the Inuit had had more than two centuries of acculturation, beginning with the Moravian Missions that had been established along the north coast as early as the 1770's. Inter-racial marriage, the pervasive use of English and corresponding diminution of Inuktitut, and the introduction of the wage economy in places like the Nain fish-packing plant— stuttering since the disappearance of the cod fishery—left the impression that the Inuit were well on their way toward a successful synthesis of an Inuit-Euro-Canadian way of life. Then too, there was the Inuit-owned labradorite quarry development at Ten Mile

Bay, just outside Nain. They had clearly mastered the use of high explosives and the handling of great weights all on their own, not to mention the management of the operation as well. Such enterprise and expertise would have made some of the Inuit, one would have thought, prime candidates for mining jobs at Voisey's Bay.

The board of directors of the LIA endorsed the prototype AIP, but as the meetings with the full LIA membership got underway in Nain, it became clear that the agreement was in for a rocky ride. The meetings were not open to the news media, but eventually someone, presumably a disgruntled LIA member, leaked the six-page summary of the agreement to the press, and soon fax machines across the country, including my own, were humming with the specifics of Canada's latest comprehensive land claims settlement offer.

At first glance, the numbers in the proposed AIP appeared impressive. The Labrador Inuit would receive surface rights to some 72,519 square kilometres, an area the size of the province of New Brunswick. Included in this Labrador Inuit Settlement Area or LISA would be a National Park of 7769 square kilometres, and an offshore area comprising 44,197 square kilometres.

Included in LISA would be some 15,798 square kilometres, to be designated Labrador Inuit Lands (or LIL), which would be owned outright by the Inuit as Category One lands. The Inuit would, however, share sub-surface resource (mining, oil, natural gas and quarrying) revenues on LIL with the provincial government. The Inuit would receive 25 percent of such annual revenues, the government 75 percent. IBAs would be required for all developments on LIL, and for all major developments on LISA. The Inuit would also be entitled to a 5 percent share of provincial mining tax revenues on all projects worth more than $40 million within LISA. The total of such revenues would be capped, however, at "no more than the average Canadian per capita income."

Most of the criticism of the agreement within the closed-door meetings in Nain, it was rumoured, concerned the size of LIL, the land the Inuit would own outright. Some LIA members contended that the area was much too small, representing as it did only a fraction of traditional Inuit lands and waters in Northern Labrador. (It was none of my business, of course, but I couldn't help wondering if the Labrador Inuit were aware of the extraordinary degree of control their Québec cousins had successfully negotiated through the Raglan IBA on lands where their title might have been weaker than the Labrador Inuit's on LISA's?)

But the largest part of the memorandum under discussion was devoted to Inuit self-government. And the largest part of that section was devoted to guaranteeing the rights of the minority, i.e. non-Inuit, residents of Northern Labrador. This has long been a concern of constitution framers in democratic governments, dating from the Federalist Papers, which had preceded the establishment of the Constitution of the United States. The American Founding Fathers were concerned, among other things, with the rights of the wealthy, property-owing minority in a then-untried system in which the majority of voters would be neither rich nor large landholders (nor, of course, would they be slaves, women or aboriginals, who would not be accorded the right to vote until much later).

In the case of the LIA proto-AIP, negotiators for Ottawa and the province were clearly at pains to protect the rights of "non-beneficiaries," in other words non-members of the LIA, who might eventually take up residence in Northern Labrador. The attention paid to this detail was, to me, a signal that talks about Inuit self-government had become serious, indeed, and that some form of new jurisdiction was about to be created out of Northern Labrador. "The Canadian Charter of Rights and Freedoms (will) apply to the Inuit government," the parties agreed. "Best efforts" would be made to provide Inuit jurisdiction "over health, education, welfare, child and family services, administration of justice and will and estates." Most intriguing of all, however, was a provision "that Inuit government laws will be paramount over the laws of the federal and provincial governments with respect to a defined class of matters, to be agreed among the parties, which are of vital importance to Inuit."

While the devil, as ever, would doubtless lie in the as-yet-to-be negotiated detail, the memorandum seemed to portend not only a new jurisdiction within Canada, but also a fourth level of government that was not federal, provincial, or municipal. The LIA's memorandum, I suspected, would be read with the greatest interest by land claims negotiators across Canada. A new Aboriginal jurisdiction was about to be created in Canada's north, underpinned with a sizeable, resource-rich land base over which the Inuit themselves would wield considerable control. LIA negotiators were predicting that the full AIP would be signed by February or March 1998, and Sudbury lawyer Steve O'Neill predicted that a further year would then be required to hammer out the complete agreement and to have it ratified by the full LIA membership. Even so, that would mean that a key piece of the Voisey's Bay puzzle could fall into place as early as 1999, and I began to daydream about a phone call to my friend Fran Williams, the executive director of the OkâlaKatigêt Communications Society in Nain. I'd congratulate her on . . . what? On her people having ceded the better part of the land they had controlled since time immemorial? In exchange for . . . what? Control, sort of, of part of the land over which they had exercised absolute sovereignty, before the coming of the white man? And I began to grasp the existential dilemma of all land claims settlements, from the Aboriginal point-of-view: at its heart, any comprehensive land claim settlement demands surrender of huge tracts of land over which one's forebears had "always" enjoyed dominion. No matter how large the cash settlement, how generous the land base, or how sweeping the provisions for self-government, at its core the settlement process still meant a form of agonizing surrender to the surrounding Euro-Canadian majority.

In the event, I need not have worried about my fantasy phone call, because the LIA land claim timetable would once again be derailed. As seemed to happen so often at Voisey's Bay, concrete progress toward development was about to be sideswiped out of the blue; this time by a bus called history.

13

Sui Generis (2)

O N DECEMBER 11th, 1997, the Supreme Court of Canada handed down what may well be the most important decision in its history, a judgment that could, over the next 20 or 30 years, drastically change the nature of the Canadian Confederation. The Court's ruling in *Delgamuukw* (pronounced DEL guh muk) *v. B.C.* has far-reaching implications, to put it mildly, and the mainstream news media (with the notable exception of the *Globe's* Jeffrey Simpson) were unaccountably slow to cover this amazing, unfolding legal story.

The immediate legal issue before the Court was a brace of B.C. Supreme and Appeal Court rulings on the claims of the Gitksan and Wet'suwet'en First Nations to 58,000 square kilometres of land in Northern British Columbia. Like the Inuit and Innu of Labrador, neither group had ever signed a treaty with the Crown extinguishing the right to their respective homelands.

The B.C. Justices, both at trial and appeal, rejected the Aboriginal claims. But Canada's Supreme Court Justices, the highest court in the land, overturned their rejection, administering what can only be described as a trip to the judicial woodshed to their B.C. colleagues in the process.

The Supreme Court ordered that a new trial be held, set down strict new rules for that proceeding, and changed the future course of Canadian history. The Court may also have thrown yet another spanner into the works of Inco's development at Voisey's Bay.

Steve O'Neill estimates that *Delgamuukw* has strengthened the bargaining position of the Inuit and Innu around Voisey's Bay by between 25 and 50 percent, provided the two Aboriginal groups can establish a *prima facie* case for Aboriginal title.

The Labrador Inuit Association appeared to be nearing an agreement-in-principle with the governments of Newfoundland and Labrador and Ottawa on a comprehensive land claim, and one of their lead negotiators predicted that the AIP could be signed as early as February or March of 1998. But that was before *Delgamuukw*. A source in Nain told me that the same negotiator was now predicting that the agreement, an essential first step before Inco can proceed with development at Voisey's, might not come before summer 1998, at the earliest.

Delgamuukw produced a not-so-subtle shift in the negotiating environment between Aboriginal groups and Ottawa, the provinces, and would-be developers; because each of the Supreme Court's decisions on Aboriginal title is more favourable than the last to Aboriginal claimants, delay, for the first time, begins to work in the favour of First Nations. *Delgamuukw* also makes it less and less likely that Voisey's can proceed without Innu and Inuit approval, meaning without land claims settlements with government and without Impact and Benefits Agreements between the Innu and Inuit, respectively, and Inco.

In theory Inco might try to go ahead, but it would undoubtedly be met with Aboriginal injunctions that could tie the project up for years, even decades. And, given recent Court judgments, the chances that the Innu and Inuit could win their claims appear greatly enhanced.

■

SUI GENERIS (2)

Central to the *Delgamuukw* decision is the question of Aboriginal title to land: what it means, how it was acquired, and how it can be surrendered. "Aboriginal title," the Justices wrote, not once, but several times, "is a right to the land itself." But it is also *sui generis*, that is, special and unique, and differs from our traditional European notions of land ownership in several ways. First of all, Aboriginal title is not a question of fee simple, that is, the common grid of land holdings on patented properties recorded in every Registry Office across the land. Aboriginal title lands belong, not to the individual members of a First Nation, but to individual First Nations as a whole. The land is held communally, and cannot be sold, alienated or surrendered to anyone except the Crown. Aboriginal title holders have the right to use the land in any way they see fit, and this right is not limited to traditional land uses. But Aboriginal title claimants may not use the land in any way that diminishes its value for future generations.

The Court's view, of course, is a reflection of the way that Canada's Aboriginal peoples have always regarded land and its animal, mineral and vegetable resources: it is communal, it is precious beyond the measure of dollars and cents, and it is a sacred trust endowed on present occupants for generations yet unborn. Moreover, the *Delgamuukw* decision significantly levelled the playing field for all future claims to Aboriginal title. The B.C. Court judges erred, the Supreme Court ruled, when they failed to give equal evidentiary weight to Aboriginal oral histories regarding land use and occupation. The simple fact is, as Canada's highest court sees it, great swatches of land in the west and far north still rightfully belong to the descendants of this country's first human inhabitants. This, according to *our* (i.e., non-Aboriginal) law, *our* history, *our* judges.

Interestingly, there has long been a *de facto* and *de jure* recognition in British common law that before the Commonwealth dominions were "discovered" and claimed for the Crown, Aboriginal people lived there. The British Crown acknowledged that these people had their own society, their own customs, culture and languages, and their own systems of administration and justice, and in some cases, because they were there first, these traditions are recognized and respected in law.

Next the Royal Proclamation of 1763 reserved "under our Sovereignty, Protection and Dominion for the use of the said Indians, all of the Lands and Territories not included within the Limits of Our said Three new governments, or within the Limits of the Territory granted to the Hudson's Bay Company . . ." (*Delgamuukw*, page 85.) Latterly, the Canadian Constitution Act of 1982 promised that "The existing aboriginal and treaty rights of the Aboriginal peoples of Canada are hereby affirmed and guaranteed."

The *Delgamuukw* decision is a weighty document, both literally and figuratively. It runs to roughly 90 pages. Double-spaced and set into type, it would fill a small book. But *Delgamuukw* isn't a patch on the legal proceedings that led up to it and that are now worth precisely nothing. The original trial required 374 days of evidence and argument (about two years on the average judicial calendar), produced 23,503 pages of trial transcript, examined 9,200 exhibits, and ended in a judicial decision that was more than 400 pages in length.

The Supreme Court Justices recorded these facts in their judgment in exquisite, almost loving detail, before tossing them into the dustbin of legal history. They did this for a reason, I think: to show how cumbersome an instrument the courts have become in resolving what should properly be a matter of cultural sensitivity, mutual respect, negotiation and compromise.

"By ordering a new trial," wrote Chief Justice Antonio Lamer, "I do not necessarily encourage the parties to proceed to litigation and settle their dispute through the courts. . . . The Crown is under a moral, if not legal, duty to enter into and conduct . . . negotiations in good faith. Ultimately, it is through negotiated settlements, with good faith and give and take on all sides, reinforced by the judgments of this Court, that we will achieve the reconciliation of the pre-existence of Aboriginal societies with the sovereignty of the Crown."

To paraphrase the late Sir Winston Churchill, it is better to jaw, jaw than to law, law. And woe betide the provincial premier who drags his feet on land claims negotiations or who fails to bargain in good faith.

Chief Justice Lamer concludes the above quote with a simple home truth aimed at all Canadians, Aboriginal and non-Aboriginal: "Let us face it, we are all here to stay."

■

Sobering stuff for Inco, no doubt, but I'm convinced that the implications of *Delgamuukw* for First Nations, and for Canada as a whole, may be even more profound.

Thanks mainly to the mineral discovery at Voisey's, the LIA's 20-year-old land claim is nearing a settlement. A leaked document, reported in the Newfoundland press, but largely ignored by the national media, revealed the outline for the impending agreement.

Besides outright ownership of a vast tract of Northern Labrador, the Labrador Inuit will receive a lesser title to most of the rest of the land, and a guaranteed percentage of revenue from any future development. In other words, Aboriginal self-government over a huge land base containing billions of dollars in potential resource revenue. This agreement could well be seen as a major breakthrough by First Nations across Canada and serve as a pattern in future negotiations for the Innu of Labrador and for Aboriginal groups in northern British Columbia, the Yukon, and the western Northwest Territories as well. Where land goes, economic benefits, potentially very great wealth, follow. And where economic power flows, political power is rarely far behind.

Private-sector companies, eager to get their hands on the resources on and under Aboriginal lands, will begin to negotiate the kind of IBAs pioneered by Falconbridge at Raglan with the Québec

Inuit, and at this point a kind of economic Darwinism will begin to take hold. Dynamic companies entering an environment where land tenure questions have been resolved (e.g., Falconbridge at Raglan, where the 20-year-old James Bay and Northern Québec Agreement has settled most land ownership questions) will prosper as the result of partnerships with Aboriginal peoples.

Resistant companies entering an environment of uncertain land tenure (e.g., Inco at Voisey's) will falter and face delays that may prove costly at best, downright ruinous at worst. The companies entering into partnerships (and they will likely include charter banks, oil and gas firms, and forestry and mining giants) will be rewarded for their cultural sensitivity and respect for Aboriginal culture and values. The resistant companies will be penalized. As a modified corporate mentality prevails, how long will it be before the captains of industry begin to encourage provincial premiers to allow Aboriginal representation at First Ministers and Constitutional conferences and at relevant meetings of lesser ministers?

(This is one of the wrinkles of *Delgamuukw*, by the way. The Supreme Court of Canada, essentially a federal body, is recognizing Aboriginal claims to Crown land that had previously been under the control not of Ottawa, but of the provinces. Ottawa's is not the ox being gored here.)

Even if there is no such corporate pressure, however, it will soon become apparent that Aboriginal control over so much land and economic wealth will make it difficult, if not impossible, to govern Canada without seeking the input of Aboriginal peoples at the highest levels of decision-making.

14

Spooning Pablum to the Bay Street Gang

AS 1997 DREW TO A CLOSE it was becoming apparent that the Voisey's Bay standoff, which might have come as a relief to Inco employees in Sudbury and Thompson, Manitoba, was going to have precisely the opposite effect. Even though they would not have to compete with cheap Voisey's Bay nickel any time soon, they were still going to have to pay for the ore body, which more and more seemed to resemble a millstone around the neck of a drowning man.

The first inkling of this came in late November, when Mike Sopko and Scott Hand called a news conference to spoon out still more pablum to the Bay Street gang in an effort to convince financial analysts, the Bay Street press (the same folks who brought you Bre-X), and nervous shareholders that aforesaid execs were still in charge, plummeting share and world nickel prices notwithstanding. The upshot of their announcement was, roughly, this: we plan to spend less money, close certain mines, reduce the workforce still further, produce *more* nickel and make *more* money.

VOISEY'S AS VORTEX

So far as I could tell, the news that Inco planned to eliminate 500 jobs in Sudbury and close four mines in Ontario Division was met with a shrug and a yawn among Inco employees. The workforce reduction (through attrition by pension) had been expected for some time, and morale inside Inco's mines and plants had hit rock bottom some time earlier. The mood seemed to have shifted from despondency to one of utter resignation.

"I go to work now and I just don't give a shit any more," one Inco mine employee told me. "Management doesn't seem to give a shit either, so why should I? I've worked for the company for over 30 years, and I've never seen things this bad." And this was several weeks *before* the Sopko-Hand news conference.

The factoid in their remarks that should have really set alarm bells ringing, both in Sudbury and in Manitoba, was Inco's changes in capital expenditures for 1998 and 1999. The company had planned to spend $560 million (all figures U.S.$) on its operations worldwide in the next year, but that figure was being reduced to $480 million, an $80 million cutback. Instead of spending $209 million in Manitoba and Ontario Divisions, we were told, the company would reinvest only $130 million. In other words $79 of the $80 million slashed from capital budgets would come out of Inco's existing Canadian operations. Similar, unspecified, cuts were planned for 1999 "as these operations pursue the strategy of maximizing their profitability and cash flow," intoned Inco's news release.

Did anyone north of the Toronto city limits believe for a New York minute that Inco would really be able to boost nickel production to 440 million pounds in 1998 from 392 million pounds this year?

At this juncture, and for the foreseeable future, Inco has only three places it can actually produce nickel: Sudbury, Thompson, and Indonesia. Production in Indonesia has been reduced, as Inco's press release conceded, because of El Niño, which produced a drought that emptied the hydroelectric dams necessary to run the Soroako smelter at anything like full capacity. The Inco executive suite may be a seat of power, but Mike Sopko can't make it rain.

And Sudbury and Thompson were going to produce more nickel with fewer people, fewer mines, and less money? Just about any Inco employee I knew could tell me a recent story about production delays due to a shortage of workers, lack of basic equipment maintenance, or both. This management policy of treating Sudbury and Thompson purely as cash cows has affected Inco production, productivity, and profitability for some time now, and it's beginning to show up on the bottom line.

And was I the only one wondering how it was that Inco eked out only a $5 million profit in the third quarter of 1997 while Falconbridge, a much smaller company, earned $24 million? Falco's Sudbury hourly rated workforce had actually risen to more than 1,400, a temporary blip to be sure, but could it be that having enough bodies to produce efficiently might have something to do with profitability? Inco's nickel production targets reminded me of the timelines for Voisey's Bay: production will start in 1998—oops make that 1999, would you believe the year 2000? Speaking of Voisey's, I mentioned alarm bells. It now appeared that roughly two-thirds of Inco's capital spending in 1998 and beyond would be earmarked for Indonesia or Voisey's Bay.

This should have served as a wake-up call, not only in the provincial legislatures of Manitoba and Ontario but also in Sudbury at Civic Square. I didn't expect a peep out of the Ontario Harris government, but it seemed high time that Sudbury's regional Council called Inco management to account. Just what were the company's mid-term plans for its Sudbury operations? It seemed patently clear that we're to be starved for ongoing capital, meaning that much-needed maintenance and upgrading may be further delayed, or cancelled altogether. This augured extremely poorly for Sudbury's long-term future. It also represented a colossal gamble on the part of Inco management, which seemed to be betting the farm that the rains would come to Indonesia and land claims would be settled in Northern Labrador, before the cash runs out. And this in a time of falling world nickel prices.

VOISEY'S AS VORTEX

The problem with this strategy was that, as never before, Inco itself had little control over the timing and processes that would determine the success of the wager. This was a company in deep, deep water. I began to suspect that Sudburians had better begin to contemplate their city's future as a world mining-nickel capital as distinct from Inco's future as a viable, ongoing corporation. And Sudbury's mines, mills, smelters and refineries can certainly still be operated at a profit. The waste, needless production bottlenecks, and lack of productivity at Inco's Ontario Division are immense, and represent a shameless squandering of a natural legacy that has served us for generations, and should serve us for generations still to come. The collective knowledge, experience, and creativity of the company's workers, staff and management in Sudbury, given the opportunity, could transform this place.

■

Cutbacks in capital spending were not the only augury casting gloom over Sudbury's Christmas season in 1997. There were also repeated rumours that Inco would shortly be placing some of its Sudbury assets up for sale. This was not the same conjecture that surrounded endless Bay Street rumours of a possible takeover of Inco itself, though given the colossal mismanagement of the company in recent years that might just prove to be the salvation of us all. There were persistent reports, emanating from Inco's corporate headquarters at Copper Cliff, that parts of Inco's Sudbury operations were on the block. They included:

- the Divisional Shops complex in Copper Cliff;
- the oxygen plant, located just north of the Copper Cliff smelter, which generates pure, highly pressurized oxygen to feed the smelter furnaces;
- the company's hydroelectric dams on the Spanish River;
- Inco's internal railway system;

- and anything else, presumably, that wasn't bolted down, didn't relate directly to the production of finished product, and could be sold for a quick profit to bolster the company's sagging bottom line.

I hadn't wanted to be knee-jerk about all this. I must confess when the rumour first reached me in late October about the possible sale of Div Shops, I wasn't sure it was such a bad thing for the community as a whole.

The sprawling shop complex, which cost $14 million and covers 140,000 square feet, was opened on August 23, 1980. Natural Resources Minister James Auld cut a pure nickel ribbon at the official opening and complimented Inco for investing in "this community and . . . this province." My, my, how times have changed. But, I reasoned, perhaps turning the place over to a smaller entrepreneur might open up the possibility of expanding production and employment, by allowing for sales outside Inco, perhaps even outside Sudbury.

Yet taken as a whole, the piecemeal sale of "non-core" assets from Inco's Ontario Division must give one pause. These assets, remember, were paid for by retained earnings generated by the parents and grandparents of those living in Sudbury. These forebears built these parts of Inco that may soon be picked apart, and which have been used to generate considerable profits ever since. The assets won't physically leave the district, of course, but wheelbarrows of cash resulting from their sale almost certainly will, destined for—you guessed it—Voisey's Bay.

When the sales are announced by Inco you can expect company flaks to put the best possible face on it by claiming the asset sales were something they planned to do all along. And it is true that when Inco made its first offer to take over Diamond Fields Resources, the-then 75 percent owner of Voisey's Bay, in March 1996, a company news release referred to "possible asset sales" as one way that the share-swap deal could be financed over time.

VOISEY'S AS VORTEX

In December 1997, Inco Sudbury spokesperson Corey McPhee announced that the company's hydroelectric dams were on the block, and a couple of months later Inco finalized a deal to sell its alloys division, parts of which had been in the Inco family for a very, very long time, pending the approval of U.S. regulators. I refrained from commenting publicly on this at the time, because I was convinced, in the end, that the sale was part of a corporate strategy to divest "non-core" (i.e., non-mining, smelting, and refining) assets in favor of a long-term commitment to Inco's core business.

The sale of whole divisions, maybe. But stripping out saleable assets *within* a division struck me as something else again. I think what we saw was a money grab by an Inco management that is increasingly desperate for cash flow just to keep the doors open and the lights on.

Inco lost money in the fourth quarter of 1997 and well into 1998. We're clearly heading into a cyclical price downturn for both nickel and copper, and Inco profitability during such periods would be severely tested at the best of times.

But now, with the Asian economic collapse (how many times has Inco boasted about its heavy exposure in Asian markets?), and the drought in Indonesia that continues to plague smelter capacity there, the options open to Inco management are increasingly circumscribed. And then there's Voisey's Bay. Factor in the ongoing exploration, legal, overhead and financing costs of an operation that won't produce a dime of revenue until some time in the next century, and you can begin to see how asset sales are one of an ever-dwindling number of ways to make ends meet.

But what of Sudbury's— and Ontario's—interest in all this? This is our patrimony, our inheritance, we're talking about here. For more than a century, the people of Sudbury have been nurtured and sustained by one of the greatest integrated mining and metallurgical complexes the world has ever seen. Are we now to stand idly by and watch it be dismantled, starved of capital investment, and run ever more deeply into the ground? A Steelworker friend of mine called recently, worried sick about the future of her union, and her company. Why, she wondered aloud, is no one speaking out about the rapidly deteriorating situation at Inco?

"If we go on strike for 28 days, there's the Chamber of Commerce in the newspapers telling everyone it's costing the business community a million dollars a day in lost sales. But why aren't they speaking out about what's happening at Inco right now? Don't they realize how serious this is?"

Sometimes corporate policy has such an overwhelming consequence for a community that policy becomes a matter of public interest. Just ask Newfoundland Premier Brian Tobin. So where *are* our business, union, municipal and provincial leaders? Why aren't they speaking out, demanding answers? I understand their preoccupation with downloading, restructuring, and the fiscal hash Ontario premier Mike Harris has made of municipal finances. But there are times when politicians must prove they really can walk and chew gum at the same time, and this, I'm afraid, is one of them.

15

Can Inco Be Trusted?

IN EARLY NOVEMBER of 1997, I received a call from a young editor in Nain. His name was Peter Evans, and he worked for a bilingual Inuktitut-English magazine called *Kinatuinamut Ilingajuk*. His publication, Evans explained, was funded by money the Inuit had received from Inco as an advance on an eventual IBA settlement on Voisey's Bay. *KI*, as it was familiarly known, was distributed throughout the north coast of Labrador and, as negotiations between the Labrador Inuit Association and Inco continued, many of his readers wanted to know more about Inco.

"A lot of people, especially some elders, are wondering who these people are, and especially whether, if an agreement is signed with them, Inco can be expected to honour its end of the agreement," Evans explained. "We thought, given the people of Sudbury's long experience with the company, you might be able to answer this question for the readers of our magazine."

I pondered the assignment and asked many of my friends in Sudbury what I thought was the key question troubling the Inuit: "Can Inco be trusted?" Mostly people just laughed, implying that the question was naive in the extreme, or that it had been answered long ago. Neither response seemed to be of much help to the Inuit. Instead, I decided to provide clues to the answer by sharing the following events from Inco's long history of operating in Sudbury.

■

In the early spring of 1902, Sudbury newspaper editor James Orr had a problem. He'd heard persistent, and reliable, rumours that his town's major employer, the Canadian Copper Company, was about to be sold. Its name notwithstanding, the Canadian Copper Co. was actually U.S.-owned and one of the world's largest producers of nickel, and Orr had heard that it was about to be bought out by the Orford Copper Company, another American-owned concern.

Orr ran a story to that effect in the March 13, 1902, edition of his *Sudbury Journal*, only to have it flatly denied by company officials. The following week, on March 20th, he dutifully printed a retraction. "No Sale!" ran the headline. "This week we are authorized by the officials of the C. C. Co. to contradict the report, and to say that there has been no transfer of stock whatever."

One can only imagine Orr's consternation, when, just 12 days later, the *New York Times* carried a story announcing the creation of a new nickel company, which would include the assets of both the Canadian Copper and Orford Copper Companies, and which was said to be closely linked to the mighty "steel trust"—the United States Steel Corporation. Thus was the International Nickel Company—Inco—born.

It was a pattern that would repeat itself again and again over the life of the newly created corporation: company officials, determining that the interests of shareholders in New York and London came before their employees and the communities that depended on them, would deny that a fact was a fact until *they* had decided to share that fact.

■

Another abiding characteristic of the new corporation was highlighted by the debate that raged for years over the refining of Canadian ores in Canada. For the first quarter-century or so of Sudbury's life as a nickel mining centre all of its smelted ore (or "matte") was shipped to the Orford Copper plant in New Jersey for refining. Canadian economic nationalists, certain politicians and newspaper editors argued that Canada should be refining its own nickel in Canada, rather than exporting it to the U.S. in a semi-finished state.

"Oh, no, no," said the officials of the American-owned Canadian Copper, Orford and International Nickel companies in so many words, "we couldn't possibly refine nickel matte in Canada and make a profit. And what's more, if you force us to do so, we'll pull up stakes in Sudbury and mine our nickel ore in New Caledonia." (Sound familiar?)

Matters came to a head during the First World War, which Canada entered in 1914, but in which the Americans remained neutral until 1917. Canadian nickel, an essential ingredient in war materiel, was shipped to the U.S. for refining, and Inco then sold it *to both sides*, so that Canadian boys were being killed by Canadian nickel mined in Sudbury and fired from German guns. Canadian public opinion was understandably outraged by this state of affairs, but the company stubbornly stood its ground until the *Deutschland* Affair of 1916. In November of that year an American newspaper reported that a German submarine, the *Deutschland*, had just left a U.S. port carrying 600 tons of refined nickel. This was too much, even for the embattled Canadian and Ontario governments, which had refused to force Inco to refine its matte in Canada. (Robert M. Thompson, the first chairman of the Inco board, boasted that he had personally paid Sir Wilfrid Laurier $5000 in stock profits that he had purchased for the prime minister.) The Royal Ontario Nickel Commission was appointed in 1916, and it concluded, in no uncertain terms, that Inco could afford to refine its nickel in Canada, and that the company was, moreover, a major despoiler of the air and water in and around Sudbury. In 1917

Inco finally established its Port Colborne, Ontario, nickel refinery, which still operates to this very day.

■

It took another world war before Inco's employees in Canada realized another long-cherished dream: the establishment of a trade union at the company's operations in Sudbury. As Sudburians knew only too well, Inco ran its mines and mills here with an iron fist. Bribes were often demanded (and paid) in return for hiring, and firing was done at company whim. Promotions often went to Anglo-Saxon employees, and the company was not above pitting one ethnic group in the workforce against another in order to divide and conquer.

Repeated attempts to unionize had been met with the infiltration of meetings, mass firings, and the blacklisting of organizers throughout the hard rock mining industry. But by 1942, a wave of industrial organizing was sweeping North America, and the formation of a local of the International Union of Mine, Mill and Smelter Workers at Sudbury's Inco operations appeared imminent. Inco responded with a last, desperate attempt to intimidate its workforce. Shortly after 5 p.m. on the afternoon of Tuesday, February 24, 1942, a dozen men entered the downtown Sudbury office of the Mine Mill union. Two union organizers were severely beaten, windows were smashed out, and an office chair, papers and files were thrown on to the sidewalk two storeys below. Two of the assailants soon confessed to the union what had happened. The thugs were Inco shift bosses from Frood mine who had been provided a perfect alibi by the company: they were "punched-in" at the mine time clock before being sent on their mission by the mine superintendent.

Union organizers quickly cranked out a leaflet documenting the incident, and 10,000 copies of it were distributed to homes across the Sudbury Basin a few nights later. Far from frightening Sudbury's workers, the attack seemed to renew their resolve. In February 1944

CAN INCO BE TRUSTED?

Mine Mill Local 598 was certified as the official bargaining agent for Inco's Sudbury workers. Over time the union fought for, and won, higher wages, better job security, improved working conditions and pensions for its members. But none of these gains came without a struggle. Strikes took place in 1958, 1966, 1969 and 1978-79. Shorter strikes were called in 1975, 1982 and 1997. Local 598 also took a lead role in discouraging bigotry and racial and ethnic discrimination, helping to make Sudbury the progressive and relatively tolerant community it is today.

■

I moved to Sudbury in the spring of 1974, and the following winter I began to hear stories of a mysterious "killer fog" that descended over the Trans-Canada Highway near Inco's operations in Copper Cliff just west of Sudbury. The fog, which occurred only in deep winter, had caused numerous multi-vehicle accidents over the years. Its source appeared to be Copper Cliff Creek, which meandered through the town of Copper Cliff and then ran parallel to the highway. Even on the coldest days the creek never froze, and both a prominent area MPP and the union that represented Inco's workers had long insisted that the water never froze because of effluent discharged into the creek from Inco's Copper Cliff operations, and that the creek should at least be covered over to prevent the fog from forming. One cold winter's night in February 1975 I received a call from an Inco pensioner: "Mickey," he said, "the fog's out tonight." I hopped into my car and drove to Copper Cliff. One minute the conditions were clear and fine, and a second later I had, without warning, driven into the densest fog I had ever seen. I could not see the hood of my own car.

I began to do research on the fog and soon learned that, because of union and political complaints, the Ontario Water Resources Commission (a forerunner of the Ontario Ministry of the Environment) had conducted a study of the creek and concluded Inco was not to blame for the fog. The following Sunday I donned cross-country skis and packed my rucksack with a household thermometer, notebook, and camera with a telephoto lens. I began skiing up the creek, back toward its source. Although the air temperature was minus 18 degrees Celsius, I discovered the water in the creek was *plus* 20 degrees. Where was such warm water coming from? I measured every tributary flowing into the creek as I worked my way upstream. At every junction I followed the warmer flow. Finally, I arrived at Inco's Copper Cliff Copper Refinery. Water apparently being discharged from this plant was 20 degrees above freezing. I believed I had found the source, both of the warm water, and of the Copper Cliff fog.

Charles Ferguson, Inco's local environmental officer, insisted "there would probably be a fog if Inco wasn't there." An enforcement officer with the Ontario Ministry of the Environment echoed the company's position: "We're just not convinced [that forcing Inco to cover the creek] would solve the problem." My story on the killer fog of Copper Cliff appeared in the *Globe & Mail* on March 14, 1975. By May I was able to report a follow-up: the Environment Ministry had conducted more studies and was changing its tune. "We are implicating the company to some degree," the same enforcement officer quoted earlier now conceded. In the fall of 1976 I received a telephone call from Inco's PR office in Copper Cliff. The company was about to begin culverting over part of the creek at a cost of more than $1 million. Would I be interested in covering the ground-breaking? In other words, was I interested in giving Inco free publicity for solving a problem it had created years before? I was not.

To my knowledge, there have been no killer fogs in Copper Cliff since the culvert was installed, and a "Watch for FOG PATCHES" sign erected on the highway in the shadow of the Inco superstack has long since been removed. And Charles Ferguson, who had earlier denied any company role in the creation of the fog, is now Inco's international vice-president in charge of the environment, health and safety. As such, he will presumably be in overall charge of environmental matters at Voisey's Bay from his office at Inco's world headquarters in Toronto.

■

So, can Inco be trusted? My answer, perhaps surprisingly, to that question is a qualified "Yes." Inco can be trusted, first of all, if the people of Labrador and Newfoundland understand that Inco is not, at the end of the day, in the business of community development, environmental protection, or even mining nickel. Inco's business, simply put, is to earn profits for its shareholders. If it can't do the latter, it will do none of the former. It is as well to expect a polar bear to act like a seal as to expect Inco to behave in any other fashion.

Inco *can* be trusted, provided the community is armed with an independent, well-informed, and critical news media; provided there is a strong and aggressive union or other organization to represent the company's workers, their families, and the immediate community; and provided a clear regulatory framework is established, and rigorously enforced, which protects the people, and the environment, while still allowing Inco to earn a profit. Then, and only then, can Inco and the people of Labrador and Newfoundland mutually benefit from the tremendous mineral wealth with which their land has been endowed. That, I believe, is the lesson the people of Sudbury have learned after dealing with Inco for nearly a century.

Voisey's Redux

February 1998

16

No Smelter, No Mine?

WHO WOULD HAVE THOUGHT that a hundred or so Newfoundlanders would turn out on a weeknight in February to hear me, an ink-stained wretch from Sudbury, give a public lecture entitled "What Can We Expect from Inco?" But so great was the thirst for information about Canada's largest nickel producer, and so acute the need for public discourse concerning the proposed Voisey's Bay smelter/refinery project in Southern Newfoundland, that a standing room only crowd was indeed on hand, including a delegation who had made the 90-minute drive from Argentia.

Argentia, on the shore of Placentia Bay, is the site selected by Inco for its $1.05 billion smelter/refinery complex. Like so much else surrounding the Voisey's Bay development, this project, too, is steeped in a chowder of controversy—Newfoundland style.

The Citizens' Mining Council of Newfoundland and Labrador has taken the smelter project, in the absence of a full environmental review, to the Federal Court of Canada; the case was scheduled to be heard in February 1998 in Vancouver.

The CMC were co-sponsors of the meeting, and they braced themselves for an earful from the Argentia delegation. Being desperate for jobs, the latter are not nearly so punctilious on environmental matters, and do not welcome, to put it mildly, the prospect of yet more delay to a project that seems to promise economic deliverance.

I attempted to avert a confrontation by pointing out that the smelter/refinery project is currently delayed, not due to some nefarious environmentalist plot, but owing to depressed nickel markets and a cash-strapped company.

"Why not seize the delay and use it as an opportunity to demand, receive, and analyze answers about the smelter's environmental impact from the Voisey's Bay Nickel Company?" I suggested, to the accompaniment of general head-bobbing. The Argentia folk, it turns out, are not opposed to any environmental review, only to "unnecessary" delays in the smelter development. Their 75 percent unemployment rate notwithstanding, "People are not willing to accept anything for the sake of jobs," asserted Ship Harbour resident Cathy Griffiths.

Conflict over the court case was a mere murmur, however, compared to the crowd's reaction to my modest proposal. Since it's clear that the entire Voisey's Bay project is going nowhere fast owing to the conflicting demands of the players, I suggested that the company be allowed to highgrade the Ovoid open pit, ship concentrate to Sudbury for smelting, and accumulate the cash with which to build a Newfoundland smelter/refinery, perhaps in ten years' time.

Well, sweet Mary, Jesus and Joseph, as the locals might say. It's one thing for Newfoundlanders to squabble amongst themselves about a court case, quite another for a Come From Away to tell them what to do with their ore. A couple of the boys quickly rounded on this Sudburian, and told him, in no uncertain terms, where he could stick that proposal. Voisey's Bay ore was not leaving Newfoundland without being smelted and refined in Newfoundland.

That was all very well, I responded gently, but just where did they propose to raise the billion bucks with which to build the thing? Still more invective followed.

"I just want to be very clear about what you're saying," I responded. "You're saying the ore stays in the ground if Inco won't or can't smelt and refine it here?"

"That's right, b'y. She stays right in the ground fer as long as it takes."

I was grinning broadly by this time. "Well, that'd make the people of Sudbury very happy."

Coincidentally or not, the very same debate blew high, wide and handsome in the Newfoundland media the next day, with Mike Sopko reportedly seeking some relief on the smelter/refinery commitment and Chuck Furey, Newfoundland's Mines and Energy Minister, cutting him no slack at all. Newfoundlanders, I now understand, are extremely wary of promises of industrial development; they've been burned so many times before, from the painful memory of the Upper Churchill development that enriched Québec but not The Rock, to the now defunct Baie Verte asbestos mine, which took an atrocious toll of miners' lives from asbestosis, to the collapse of the cod fishery. These are sweet, tender, hurting people who have seen their natural resources squandered time and again. This time, they resolved, things will be done differently, and they will be done right, and it might very well be political suicide for any Newfoundland government to permit Inco to smelt and refine offshore.

As logical as it might seem, the obvious middle ground—that Sudbury receive Voisey's Bay ore as a temporary, stop-gap measure that would allow Newfoundland and Labrador and Inco to get into the game—was a dog that just won't hunt, at least with politicians. Inco's problems at Voisey's Bay had become a whole lot worse, with no compromise in the offing. And this latest impasse could just be the deal-breaker that would put a stop to the Voisey's development, at least for Inco, and at least for the foreseeable future.

17

"We Are the Innu of Northern Labrador..."

O<small>N THE EVENING</small> of Thursday, February 5th, 1998, I sat in the village of Davis Inlet and watched, in utter astonishment, as representatives of the Voisey's Bay Nickel Company explained the company's mine and mill Environmental Impact Statement to members of the Innu Nation. The meeting, which was resolutely shunned by everyone in the community except for a handful of elders, was opened by Jackie Penny, a human resources officer from VBNC's St. John's office. "We like to open all of our meetings with a safety tip," the youthful and personable Penny explained, awaiting the translation of her words into *Innu-eimun*, still the first language for most of the people here.

"You know, it's winter outside, and when we come indoors we track in snow on our boots and it melts on the floor, and turns into water. And that makes the floor slippery. So I'd just like to caution everyone to be careful when they're walking on the floor here."

Monique Rich immediately asked a question of the translator. I took it to mean: "What did she just say?" When told, Rich made a non-committal affirmation but I, and not for the last time that night, shook my head in disbelief. Here were gathered some of the most respected elders of the Innu Nation, a people whose abilities to survive for millennia in one of the most hostile environments on the face of the planet were often described as "heroic" by anthropologists, journalists, and fur traders. And yet here was a representative of a far-away mining company who had arrived in the community only hours before offering a doubtless well-meaning, if somewhat simple-minded, "safety tip" to a group whose collective wisdom about safety and survival on their land might have inspired admiration, awe, and the most profound respect.

The moment was surreal, ironic, and absurd. "If last night's meeting is the nub of the interface between VBNC and the Innu community, the company has a long, long way to go," I wrote in my notebook the next day.

In hindsight I realize that my observation represents a considerable understatement. And I also have concluded that, given their mobility, courage, and proclivity for direct, albeit non-violent, confrontation with authority, the Mushuau band of the Innu Nation are, or should be, Inco's worst nightmare.

∎

"She was born at Voisey's Bay," the translator says of Monique Rich to Penny and her cohort Bill Napier, VBNC's vice-president of environmental affairs. "She saw her first white person there when she was a little girl and her family went there to trade and hunt and fish. She's pissed off at Voisey's Bay Nickel for coming around and starting to dig a big hole in Mother Earth. This breaks her heart. That's her home. She was born and raised there, and her father was buried there."

Through the translator other elders begin to berate the company representatives. "If the Innu people went to London and started drilling on Queen Elizabeth's backyard, we'd go to jail. It's not okay, it's a violation," says an elder named Edward. "Voisey's Bay is Innu land. We as the Innu don't talk about destroying the land. My grandfather died there and was buried there, I think it was in 1922. What if *I* started drilling where *your* grandfather is buried? When the drilling started the Innu were not consulted. Did you ever hear that our leaders said you could drill?"

Napier, who does most of the talking for the company this night, listens impassively. "We recognize the Innu claim to Voisey's Bay, that's why we're negotiating an IBA."

"I don't want no money. I don't want no money," Edward mumbles.

"We're here to protect your homeland," Napier replies.

All of this is too much for Ruby, the young Innu woman who has been hired by the company to provide translation for the evening. "I have to say I'm very uncomfortable with what I'm being asked to tell my elders," she interjects, in English.

In some ways, this meeting reminds me of the one I attended a few nights earlier in St. John's. Like Newfoundlanders, the Innu have been burned repeatedly by industrial "progress" on their lands. The development of the Upper Churchill hydroelectric project, low-level flying out of Goose Bay, and now the unwanted mineral exploration have left the Innu hurting, and wary. They are also conversant with the social, environmental and economic risks of open-pit mines, having either visited or heard from their Québec cousins, the Naskapi, about the downside of the giant Schefferville open pit iron mine. That mine was closed by Brian Mulroney, acting on behalf of the Iron Ore Company of Canada in the 1970's, and the lack of benefits to nearby Naskapi settlements is an ongoing sore point. I was even told later that the Innu had been refused membership in the pit's union, the United Steelworkers of America, presumably on racial grounds. There is a kind of karma at work here, I realized. The sins of the fathers in earlier developments that ran roughshod over the area's Aboriginal peoples will be delivered on the sons, the would-be industrial developers who come later. The bill is accumulating, and the debt will eventually have to be paid. In this case, it appears, the bill is about to be presented to the Voisey's Bay Nickel Company.

■

"How do you think it went?" Napier, who seemed abashed by the meeting, asked me when the two-hour-plus session ended.

"Well, it's the start of a dialogue," I said hopefully. "But I wonder if these people want this project under any terms."

I asked Napier about the question of burial sites, and he told me that salvage archaeologists had spent over a million dollars searching for such sites in the area, and had found none. I thought it characteristic that a company spokesman would quantify what is, after all, a serious concern for the Innu elders in dollars terms. Napier had convinced himself that since the archaeologists had not found the burials they simply weren't there. But I've done archaeology and am familiar with the terrain of Voisey's Bay. Short of running test trenches on a tightly packed grid the length and breadth of this rugged area, a single burial site would be easy to miss, I told him. And besides, maybe he was interpreting the elders' concerns a bit too literally.

"It's pretty hard, in this day and age, to get a project like this off the ground without the support of the local community," Napier observed. It was the most percipient and forthright statement I had ever heard from an Inco employee about the realities of Voisey's Bay. "But there just isn't much trust there," he concluded soberly.

■

I awoke the next morning with the words of Monique Rich ringing in my ears. She had been born at Voisey's Bay, she said, and her father had died and was buried there. Here was the elder I must take tobacco to, and respectfully request an interview. Of course, nothing is ever quite that simple in Utshimassits. First, I'd need to hire a

"WE ARE THE INNU OF NORTHERN LABRADOR..."

translator, and on this morning the cold was bitter; -75 degrees, with the wind chill, somebody said. It was even colder on the back of Christine Cleghorn's snow machine as she gave me a guided tour of the village.

Snow machines are the sole means of mechanized conveyance here in the winter, and they also move freight, pulling boxy wooden sledges with a strip of iron attached to the runners. These latter contrivances are called komatiks. Christine, whose official title is Voisey's Bay Assessment Coordinator for the Innu Nation, is an Akeneshau, or white person, but she has clearly come to feel at home since taking up her post here nearly a year ago.

She takes me to the site where the house fire had claimed the lives of those six children back in February 1992. It's on a hill overlooking the Inlet, and the place is marked with a green cross. Then we're off to the home of George Gregoire, to see if he'll translate when I talk to Monique. George is a slender, shy man in his late 40's or early 50's. An old wood stove throws ample heat. Most people here still heat with wood, and most have no running water, though they do have cable TV bringing them the latest news from Detroit. The house is sparsely furnished, and I am struck by the lack of amenities, of comfort. Is it due to poverty? Indifference? I wonder if the truth is that these are people who, until very recently, ranged far and wide on the land, carrying everything they needed on their backs. Their thinking is still geared to the essentials required for survival, and not to the material "stuff" that at once makes comfortable, but also greatly encumbers, our own daily lives.

We chat awhile, and the conversation comes eventually to the subject of gas sniffing among the young people of Utshimassits. Although conditions in the village have improved since 1992, a substance abuse counsellor had earlier told me that Utshimassits still had 107 "sniffers" as of February 1998.

I ask George the cause. "Well, it's not because they're not getting what they need from their parents, I don't think. But then again maybe it is."

George's son Pete, a handsome young man of perhaps 18, appears from a bedroom and George asks him something in *Innu-eimun*. To my surprise Pete suddenly becomes tearfully emotional as he replies.

"He says it's because it makes him feel powerful, like a shaman," George translates. I interpret this to mean that it makes him high, providing momentary escape from this isolated community, so lacking in the material things that bombard the Innu youth on cable TV. My heart goes out to Pete and the others of his generation, caught as they are between the traditional Innu lifestyle on the one hand, and the slick blandishments of television on the other. Two totally alien cultures are colliding here, and it is the young people who are stuck at ground zero.

■

A few minutes later Monique and George appear at the door of the house where I'm staying. I hastily brew a pot of tea, and we sit down at the table. I start the interview by asking Monique her impression of last night's meeting. She is critical first of all, not of the company, but of her own people.

"She didn't like to see one of the elders drunk," George translates, "but there were many things she didn't like, she didn't like what she heard." I understand this at once—one of the most critical, and voluble, of the elders had clearly been drinking.

She is 59, Monique says, and yes, as a young girl she had lived an entirely traditional lifestyle. The family had followed the caribou, she recalled, moving inland as far west as the George River, which is near the Labrador-Québec border.

"I remember when the Innu had no white man's clothing, only deer skins. We dressed warmly in caribou skins, and when it's really cold we didn't move around, we stayed in one place. Caribou skin was also used for canvas, to cover our tents. We had three or four different kinds of tents.

"WE ARE THE INNU OF NORTHERN LABRADOR..."

"I had heard about white people. But who are these people—are they human beings? My parents went to trade near Nain, and that was the first time I had ever seen a white person. I was surprised. 'Why do they look so different?' I wondered.

"I left Voisey Bay when I was nine, and I never went back until the protest there last summer. I was very upset, not just by looking at the mining camp, but I remembered so many things in the past. The family members, they're all gone now, except for two. . . . Maybe the company thinks we are all gone. But some of us are still here.

"I think about these things. Why didn't the government provide us with food? Now the government is trying to steal the land away. In the early days, before welfare, food came from the country, not from cans. It was fresh. Ever since I can remember we always had enough food for the winter."

I'd heard that many Innu do not consider themselves Canadian. I ask Monique if she does, and it takes George a while to translate the concept.

Monique laughs finally. "No, I don't feel Canadian. I'm a true Innu of this land. All of my brothers and sisters were born in the country. None of them were born in the hospital. . . . I love my land."

Over and over again, Monique expresses worry over the environmental impact of mining on the Innu land. "Just yesterday, some young men went to Voisey's Bay to get fresh fish—char, lake trout. They do this every year. But in three years what will be left there?"

But the Innu land is a huge place, I pointed out. Couldn't the people go somewhere else?

"Yes, there are many places. But these are rivers where the people go every spring. The government of Canada should know—we are the Innu of Northern Labrador. We are very poor. We don't have the kind of money the white people have. We don't have our own planes, and yet the government says, 'We want this land.'

"Look at this community now. Maybe the provincial government thinks they did a lot for the Innu when they built the first house? We still have no water. Right now, I have no wood to burn. It's cold.... Since we moved to this island me and my late husband never got a new house here. Where I live now is my father-in-law's."

But if she doesn't consider herself to be a part of Canada, why should the Canadian government do anything for her? I wonder.

Monique looks at me hard for a moment, then says something with a dismissive gesture. I eagerly await George's translation.

"She don't care. She never got no new house, anyways."

The three of us dissolve in laughter.

18

Disarmed and Dangerous: Katie Rich on Home Ground

"GET AWAY FROM THERE! I mean it!" There was, suddenly, a harshly convincing edge in the voice of Christine Cleghorn as she tried to shoo away two ten-year-old boys who were hovering around the back of her parked snow machine.

It took a minute, but then I understood: Cleghorn, not a mother herself, was afraid the boys would grab on to the snowmobile as we drove away, catching an illicit—and breathtakingly dangerous—ride on their bellies as we sped through the snowy streets of Davis Inlet.

I'd seen the Innu children pull such stunts before; jumping off the roof of the school into the snowbanks below and, when that paled, diving head first and doing full somersaults in the air before landing. It was hard not to be impressed with their fearless, nimble grace, but there is a videotape of a different side of this daring: the kids of Davis taunting and teasing the RCMP, running circles around them, while committing devilish and joyful acts of vandalism to the property of the Voisey's Bay Nickel Company.

The tapes were disturbing and intriguing, vaguely reminiscent of evening news footage of the youths of Belfast or the West Bank. But this was not some troubled global hot spot. The video had been shot in late August 1997 in Canada or, as some would argue, in a sovereign territory surrounded on three sides by Canada and on the fourth by the Labrador Sea—Nitassinan, the homeland of the Innu Nation. Whichever, by coincidence of history, geography and geology, the 1,700 Innu of Northern Labrador are fast emerging as a political, economic and legal force to be reckoned with, and they are presided over by Katie Rich, who at this moment may just be the most powerful woman in Canada that most people have never heard of.

■

We sped off on Christine's snow machine, her "anger" having warded off our pursuers. It was through Christine, a WASPY 20–ish graduate student on leave from Guelph University, that I had negotiated my visit with Katie Rich on her home ground, which is, as they say, a long way from downtown. Davis Inlet is a two-and-a-half-hour flight by Twin Otter north of Goose Bay, itself a two-hour flight north of St. John's. Christine had assured me that Rich would, in fact, share some of her time in early February, if I made the trip. It was also Cleghorn who had provided me with some of Rich's vital statistics: she was a 38-year-old mother of six and grandmother of two, she had been arrested several times, and jailed once. Rich kept her official conviction framed and proudly displayed on the otherwise bare walls of her Innu Nation office. Several non-Innu men had told me that they had heard Katie Rich speak to audiences of battle-hardened business types and that she had quickly reduced them to a rapt and respectful silence. But I had never met her, and I was more than a little disappointed when, upon my arrival in Davis, Cleghorn informed me that Rich was recovering from a flu that had degenerated into pneumonia. Hopefully, before I was due to depart...

■

I was intrigued by Rich, not least because of the way that her bargaining power on behalf of the Innu had been so greatly enhanced by two seemingly unrelated events in recent months. The first, of course, was the *Delgamuukw* Supreme Court decision, which appeared to be tailor-made for the Innu.

But in mid-December 1977, even as the Court was handing down its decision on Aboriginal title, rumours began to swirl around Newfoundland of a second Labrador megaproject that would dwarf even Voisey's Bay. Newfoundland Premier Brian Tobin and Québec Premier Lucien Bouchard, Water Street scuttlebutt had it, were engaged in secret negotiations to develop a huge hydroelectric project on the Lower Churchill River, a $12 billion development that would generate 3,000 megawatts of power, making it the world's largest hydroelectric project after the massive, and controversial, Three Gorges Dam on China's Yangtze River. Some of the electricity generated would be transmitted, via a $2 billion underwater cable, to the island of Newfoundland to power the Inco smelter/refinery.

Many Newfoundlanders believed the Lower Churchill was yet another piece of Brian Tobin's grand vision to galvanize Canada's poorest province. Besides oil and natural gas from the offshore Hibernia and Terra Nova fields, Newfoundland and Labrador would soon become an exporter of base metals and hydroelectric power as well. Political pundits, including the *Globe & Mail's* Hugh Winsor, speculated that, once Tobin had completed this wondrous transformation of Newfoundland and Labrador he would use the accomplishment to vault into a serious run for the leadership of the federal Liberal Party and, eventually, the prime ministership of Canada.

There was at least one serious hitch to this grand design, though the news media rarely mentioned it: like the Voisey's Bay deposit, the proposed Lower Churchill dam site lay squarely on Innu land, a scant 100 kilometres upriver from the Innu community of Sheshatshiu. By early 1998, armed with the *Delgamuukw* decision, backed by elders like her mother, and children like those in Davis, Katie Rich had quietly emerged as the only negotiator in Canada with her adversary's $14 billion, and political future, to play as a bargaining chip.

■

My glasses were badly fogged as I stumbled out of the perishing February wind into the headquarters of the Innu Nation on the outskirts of Davis Inlet. I was due to leave that afternoon and, fearful that Katie Rich was still too ill to keep our appointment, I headed directly for the unprepossessing office of the president of the Innu Nation. I was disappointed to find only a diminutive, bespectacled woman clad in a plain white cotton sweatshirt focusing intently on a laptop computer. Too young, I thought, and too, well, *ordinary* looking to be the great Katie Rich. I retreated to an outer office to warm my spectacles and cool my heels.

But after a few minutes the woman behind the desk waved me in and pointed to a chair. Hoping my confusion was not evident, I began with the same question I had posed to her mother the day before: given her inordinate leverage over events in Canada, did Katie Rich consider herself to be a Canadian?

"Never," she replied emphatically. "I've said that before. Look back at what's been done to the Innu. When Newfoundland joined Canada in 1949 were the Innu consulted? Of course not. This community has been relocated three or four times. Were the Innu consulted? Of course not. The government uses the term 'What's best, what's best for the Innu,' without asking us. The Innu have been shipped here, and shipped there, like cattle. Another thing Canada gave us was the Church, and the first thing the priests translated into the Innu language was the Bible, but nothing to teach our children their own language. So, no, I have very little interest in ever becoming a Canadian."

I asked her about the conviction notice hanging on the wall. It stemmed from a December 1993 incident when she was the chief of Davis, Rich explained. "The court party was in town and the judge kept sentencing people to jail, sending them out of the community. This was really bad, so I called two of my friends. We had a meeting and discussed two options: we could go into the courtroom and disrupt the proceedings, or we could go into the courtroom and attack the judge." Given their devotion to non-violence, I wasn't surprised when Rich told me the former strategy had prevailed. The trio hastily gathered the women and children of Davis, marched into the courtroom, and removed the judge from the bench. "Word spread like wild fire as to what we were doing, and soon there was a crowd outside the building," Rich recalled with a slight smile. The crowd surrounded the court party and escorted its members to the air strip and on to their waiting plane. Shortly afterward, the Labrador Superintendent of the RCMP came to see Rich. "Do you want us to leave, too?" he asked her.

"If you want," she nodded. It would be three years before the RCMP returned to the community, and then only under the terms of a policing agreement with the Mounties that Rich negotiated herself. This egregious rejection of provincial control was anathema in St. John's, however, and the following summer Newfoundland Justice Minister Ed Roberts ordered riot gear-equipped RCMP and Canadian Forces personnel to escort a provincial judge back into the community. The Innu hastily barricaded the airstrip so that no aircraft could land, and what Rich later termed "a military invasion" never took place. Eventually Rich and her two co-instigators, one a social worker and justice of the peace, another an Aboriginal constable, served ten days in the Goose Bay jail. "We were locked up together. We had a great time," Rich said of the experience.

Throughout our conversation Rich spoke simply and directly, and her demeanor was anything but intimidating. Yet beneath this unassuming exterior, lay a will of steel and a sense of self-confidence bordering on cockiness. Rich's in-your-face negotiating style has become the stuff of legend. The Voisey's Bay Nickel Company had leased some of downtown St. John's poshest office space for its headquarters and one of her favourite moments, Rich told me, was lighting up a cigarette once she and her suit-wearing adversaries were comfortably ensconced in the hermetically sealed boardroom where, of course, no smoking was allowed.

Innu environmental advisor Larry Innes was an occasional participant in Innu talks with Inco or the provincial or federal governments, and usually, he noted, Rich and Cleghorn were the only women at the table, a gender imbalance that produced an ambiance, in Innes's words, "of a whole lotta dick waving." Innes remembers once watching Rich sizing up her opponents before establishing eye contact with the one who seemed most full of himself. "Sorry, what's your name?" she asked, leaning across the table "and just what do you do here, anyway?" "You could almost hear the air rushing out of the stuffed shirt," Innes said with a chuckle.

Although she punctuated almost every sentence with a smile or a good-natured laugh, Rich, who was elected leader of the entire Innu Nation in 1997, could barely disguise her contempt for the men of Inco and its wholly owned subsidiary, the Voisey's Bay Nickel Company.

"A lot of people have been turned off by what's happening at Voisey's Bay and they feel these people simply can't be trusted," said the soft-spoken Rich, and this time there was no smile. "Did you hear about Inco's offer of 'gifts'?" she asked me. I tell her I had not. "Last summer, before the protest, the company offered to build a treatment centre here in our community.

"'If you're willing to give gifts, then there's certain things you have to do,' I told them. 'First, declare a six month shutdown on work at Voisey's Bay. Then come and stay in our community for six months and ask the peoples' permission for what it is you want to do. You'll be surprised how people will react if you ask permission.'

"Six months, huh?" came the reply from the Inco pooh-bah, the way Rich told it. "How about three months? Or one month? Could we make it a week?"

"I told Inco to keep its money," Rich remembered with a smile. "Stew Gendron called me, very upset. And Mike Sopko also, very upset that we wouldn't accept their money."

I couldn't help thinking about the millions of dollars in "gifts" that Inco has bestowed on the people of Sudbury, helping to build everything from Laurentian University to Science North, and how our local politicians now refuse to "point fingers" at the company: even though they must know, as many of us are beginning to fear, that if the unthinkable happens and Inco sinks in the cruel waters of the Labrador Sea the company is likely to pull much of Sudbury down with it in the aftertow.

The morning has passed quickly and I ask Rich, in conclusion, about her vision for the future of the Innu Nation. "You know, fifteen years ago everyone would have said, 'Davis is hopeless.' Everyone was drunk, there were problems of child abuse. We thought having a house without heat, without running water, without power, was normal. But then we called community meetings and began to talk openly, and for the first time, about our problems. Now I see more and more young people staying sober, making the changes they want to make. We were once a strong people, a strong nation. We will be like that once more."

■

It was lunchtime and I asked Rich if she'd give me a lift back into the village. We were flying along on her snow machine when we spotted two young boys standing in the road, and what happened next happened very, very fast. Rich flicked the throttle, a gesture that was both an acknowledgment and an invitation, and the sled slowed slightly. As we came abreast of them, the boys suddenly disappeared. I turned around and saw the first boy holding on for dear, joyous life to the back of the snowmobile, the second clinging happily to the ankles of the first as the president of the Innu Nation, with supreme, dexetrous confidence, towed the future of her people into the centre of town.

■

The next time I heard Katie Rich's voice was Monday, March 9th, on the CBC Radio news. The Innu had travelled to Churchill Falls to disrupt what was to have been a carefully delivered Lower Churchill Falls development announcement by premiers Tobin and Bouchard. A road barricade stopped the two premiers dead in their tracks, and the ensuing havoc wreaked by the Innu forced the cancellation of live television coverage of the great event. It was all very confusing, but it appeared that the RCMP got the bright idea of locking most of the protestors, who were occupying the media centre, inside the building.

Some of the Innu managed to elude captivity and continued to harass the premiers anyway, but the lock-down itself had several consequences that were, apparently, unforseen. The Innu were locked up together with the news media and the satellite uplinks and TV lights and studio cameras intended for the premiers, thus facilitating a hijacking of the media agenda. They were also locked in with a sumptuous buffet laid on for the various political staff and reporters, not to mention a bank of free phones installed to allow reporters to file their stories. It was reported that the Innu thoroughly enjoyed the buffet and had a fine time phoning their friends, on the government's dime, "from coast to coast to coast." While the assembled politicos and utility boffins were busy handing out business cards with telephone and fax numbers to reporters, the Innu focused on a single bite of information: the address of the highly sophisticated Innu Nation website at www.innu.ca.

I listened to Katie Rich's calm, clear voice on the radio for a few seconds, but then my attention was drawn to the sounds of what was going on around her. She was clearly surrounded by kids, laughing and in high spirits, who had just helped their parents and grandparents publicly humiliate two of the most ambitious men in Canadian politics. It was, I reflected later, a clear case of the background imparting more information than the foreground. I'm sure most listeners missed the significance, because it sounded for all the world like nothing more than a group of children—at play.

19

Notes from the North Coast

"Freedom of the press is guaranteed only to those who own one."
– A.J. Liebling, 1960

IT HASN'T TAKEN LONG for editors on the coast of Labrador to discover the iron fist inside the velvet glove of the Voisey's Bay Nickel Company. When I visited in October 1996 I was tickled to see published photos of VBNC executives presenting cheques to a wide variety of worthwhile causes, ranging from the Labrador Winter Games to a new health centre in Goose Bay-Happy Valley. Needless to say, it all reminded me very much of Sudbury. Corporate largesse also extended to the Labrador media, but, as with so many other of its goodwill-building initiatives in this exquisitely beautiful and sensitive land, Inco got more than it bargained for, and thereby hangs a tale.

The story begins with an institution that long predates Inco's presence on the coast, the Nain-based OkâlaKatigêt Society. Operating on shoestring funding from the federal government, the OK Society, as it is commonly known, provides 20 hours of weekly radio programming in Inuktitut and English through a network of repeater transmitters to a handful of communities up and down the Labrador coast. The Society's staff of 13 also produces four half-hour television programs monthly that are shown across the Canadian Arctic on the Television Northern Canada, or TVNC, network. It's no exaggeration to say that the OK Society is the voice, as well as the eyes and ears, of the Labrador Inuit people. And, like responsible news media everywhere, the Society has been known to run afoul of the politicos who head the Labrador Inuit Association, which will soon be running self-government in this part of the world.

Away back in the 1970's the LIA started publishing *KI*, at the time, a mimeographed newsletter, which the OK Society took over after its founding in 1982. *KI* gradually evolved into a full-fledged magazine, complete with glossy cover and newsprint pages, but it was closed by the OK Society board in 1994 due to a lack of funding. Then came the discovery of the Voisey's Bay deposit and Inco's purchase of the property in August 1996. The LIA entered negotiations with Inco over an Impact and Benefits Agreement, which will provide, among other things, payment to the LIA in return for permission to develop the resource.

LIA functionary Chesley Anderson approached OK Society executive director Fran Williams with a proposition: why not revive the *KI* magazine with the pre-IBA settlement funding the LIA was negotiating from the company? The bilingual Inuktitut and English magazine could then provide important information to Inuit readers about the Voisey's Bay mine/mill development.

Williams consulted her board of directors, who approved the application, with the proviso that neither the LIA nor VBNC would have any control over the editorial content of the new magazine, and in March 1997 Williams learned that funding had been approved for one year to produce four issues of *KI*. The OK Society advertised for an editor for the new publication, and by June had hired the 24-year-old Peter Evans, fresh out of King's College Journalism School in Halifax. He moved to Nain in June, and by late

September the first issue of *KI* was on the streets of the scattered communities of the Labrador coast.

Its contents must have been a shock to VBNC executives at corporate headquarters in far-off St. John's. Included was a full-page advertisement from the United Steelworkers Humanity Fund, a five-page photo spread of the August protest at the Voisey's Bay mine site by Labrador Innu and Inuit, and a brief interview with writer Farley Mowat, who visited Nain in September. "What I want to talk about is the absolute necessity of the people of this land to resist with everything they've got, including their blood, the gross commercial greed of the corporate world," advised the celebrated scribe, with reference to the Voisey's Bay development.

But the centrepiece of the issue was a story by Evans describing how, through a complex web of joint venture partnerships, the LIA had found itself in an embarrassing position during the August protest and court case that eventually scuttled the Voisey's Bay development, pending the completion of a full environmental assessment. It seems that, through its joint-venture companies doing work at the site, LIA members were actually employed cutting down trees for the very airstrip the LIA leadership was at such pains to protest against. Evans's story, a first-rate piece of investigative journalism, must have been as unpopular in LIA headquarters in Nain as it was at VBNC in St. John's, where the money to publish all this originated.

Popular reaction to his first offering was positive, Evans recalls, but the silence from the LIA and VBNC was deafening. None the wiser, Evans began to put together the second issue of his quarterly, hoping to have it out by Christmas. But a funny thing happened on the way to publication: the second payment from VBNC for *KI* magazine didn't arrive. Williams and Evans began to phone St. John's, where they were passed from one VBNC vice-president to another. The company, it soon became clear, was not pleased.

VBNC execs, while conceding they had no control over editorial content, complained that the publication lacked "balance" and had not attempted to impart "appropriate" information on the Voisey's Bay development. Still, Williams and Evans were told in mid-December, "a cheque was in the mail." Evans's second issue languished on the production table, the material becoming staler with each passing day. Christmas came and went, and still no cheque from VBNC. Finally, the OK Society decided to go public about VBNC's apparent welshing, reporting the story on OK Radio before giving it to CBC Radio, which broadcast the news throughout Newfoundland and Labrador. It was the talk of St. John's when I arrived there in early February.

To its credit, VBNC had cut the cheque before the story broke, but no one at the OK Society knew it at the time. So, in a hapless fashion that is becoming increasingly characteristic of VBNC management, the company wound up with the worst of both worlds: expending money to support a publication that is sometimes critical of the company, and with yet another public relations headache in Newfoundland and Labrador.

"I really don't care if they stop funding us," says Williams in reply to a visitor's query as to whether it isn't somewhat naive to expect VBNC to pay for a publication it can't control and that serves as yet another platform for anti-development sentiment. "It's not as if VBNC isn't getting something in return for the Impact and Benefits Agreement," adds Evans. By that, of course, he means access to the ore body, which is on LIA land. "I thought this was about good corporate citizenship. . . . One way or another we'll keep publishing *KI*. I'm not leaving Nain."

■

I was struck by the fact that several younger people in Nain, who had been mildly in favour of the Inco development at Voisey's at the time of my earlier visit, had now become adamant opponents of same. I was also struck by Inco's growing unpopularity on the Labrador coast. A couple of incidents may help explain why:

- An Inuit elder from Nain goes hunting in the Voisey's Bay area, just as his people have done for thousands of years. He is met by a company security guard who confiscates his rifle and sends him on his way. You can understand why company officials, with diamond drill crews operating in the bush, would be nervous about hunters packing high-powered rifles in the area. But you can also imagine how this news played when the hunter returned to Nain.
- A certain diamond drill company, subcontractors to Inco, needs snow machines and ATV's for its employees at Voisey's. Does it buy the equipment in Nain, Goose Bay, or even Newfoundland? Nope. Word on the coast has it that the company bought its equipment in Ontario.
- Nor is Inco's unpopularity limited to the Aboriginal people of the coast. A Goose Bay-based songwriter named Kirk Lethbridge is at work on a new song about the company, wherein he explains what I-N-C-O really stands for: "I Never Caught On."
- There is a strong sense in Labrador, and I heard this over and over again, that the "adjacency principle"—the notion that coastal communities, which will bear the environmental and socio-environmental impacts of the open-pit mine and mill, should therefore benefit the most from the project—is a crock.

Rightly or wrongly, the feeling is that most of the benefits, including jobs, are flowing to non-Labradorians. Even so, the Town Council of Goose Bay–Happy Valley passed a resolution and sent a delegation to St. John's to urge the provincial government to let Inco begin mining and milling at Voisey's ASAP, while allowing them to ship concentrate elsewhere.

This idea—mine now and smelt in Newfoundland later—is a non-starter on the island, whose voters outnumber Labradorians ten to one. While it would doubtless benefit the Goose Bay-Happy Valley economy, it would do little for Newfoundland, except bring a world of trouble for the Tobin government when it seeks re-election.

My impression is that Inco and VBNC are losing touch with the people of Labrador, with the possible exception of the business community in Goose Bay. But is it any wonder? The corporate execs are comfortably ensconced in push headquarters in downtown St. John's. Inco's producing operations, like Sudbury and Thompson, Manitoba, were supporting no fewer than six vice-presidents at the Voisey's Bay Nickel Company, at last count.

It takes as long to fly from St. John's to Voisey's Bay as it does to fly from Toronto to Vancouver by jet. And, believe you me, the climatic, cultural, and environmental differences between St. John's and the North Labrador coast are at least as great as the differences between Vancouver and Toronto.

Labradorians are extremely wary of Newfoundlanders, and 90 percent of the Newfoundlanders I talked to in St. John's have never been to Labrador. The people of Labrador (and this throws me every time I go there) aren't even in the same time zone as their Newfoundland cousins. The latter are an hour-and-a-half ahead of Eastern Standard Time, while the former are on Atlantic Time, the same time zone as the Maritime Provinces.

■

When I arrived in Davis Inlet-Utshimassits I thought the community was in the midst of a shed-building boom. Outside almost every house was a crew of Innu carpenters, building plywood sheds. Turns out these 16 by 20 foot structures are Davis-style granny flats. The community is bursting at the seams, and overcrowding is a serious problem, with up to 20 people living in a single, three-bedroom bungalow.

This is not the result, for once, of federal, bureaucratic neglect, but rather it is because the entire community is relocating to a place called Natuashish, on Sango Bay. It seems the feds have elected not to build additional houses in the old community, which won't exist in a few years, anyway.

The $85 million relocation project, now well under way, has created a mini-economic boom in Davis, and unlike most Aboriginal communities in the North, where unemployment frequently runs as high as 90 percent, almost everyone who wants to work is now gainfully employed.

The relocation project will take another five or six years to complete, I was told, and is yet another reason that the Innu aren't rushing to conclude deals on the Voisey's Bay project. The population of Davis Inlet, just 700 people—half of them under the age of 18—has its hands full with relocation, and doesn't need the work a nickel mine might provide.

I asked Katie Rich when—or if—the Innu would ever be ready to support Inco's grand dreams for Voisey's Bay.

"Maybe," she allowed at last. "In about 20 years or so."

Given the state of affairs on the North Labrador coast, the Innu's timetable could turn out to be closer to the eventual reality than Inco's.

Epilogue

August 1998

∎

20

Exit Inco?

"I can only urge you to be able to innovate, to roll with the punches—and there will be punches—and to strive for perfection. The single biggest hurdle you'll face as you cross this threshold today is the speed of change in our society. No person can stay on top of the growth of knowledge any longer. But we can, and must, keep our minds open to new ways of doing things, and to new ideas."
– Inco CEO Mike Sopko to Laurentian University graduates, June 2nd, 1995

During my first trip to Voisey's Bay I became aware of a peculiar temporal phenomenon. I noticed an odd, and increasing, gap developing between predictions of when a particular aspect of the Voisey's development would be completed versus the passage of real time. Over a two-month period, for example, the time required to achieve a goal would have increased by four months. I ascribed this trend to temporary conditions that would reverse themselves forthwith. Still, it was true that, so long as the two timelines continued to diverge, the completion of Voisey's would become ever more distant.

EPILOGUE

On my second trip to Voisey's, in February 1998, I was surprised to discover that, far from being reconciled, the tendency of the two lines to diverge to infinity had continued. Huge questions (where the smelter/refinery in Argentia would get its power, for example) remained unanswered, while questions whose answers had once seemed clear threatened to be reopened (a stubborn, and vocal, movement in Labrador insisted that the smelter/refinery location should be revisited, for example, in favour of a Labrador site).

By late summer 1998 the standoff at Voisey's Bay, which had begun with the Innu-Inuit protests and court case of the previous year, had become a more or less permanent condition. The immediate cause of this stasis was the breakdown, in late July, in negotiations between the Tobin government and Inco over the planned smelter/refinery project at Placentia Bay. Given plummeting world nickel prices and its own beleaguered financial state, Inco contended that the $1.05 billion smelter/refinery no longer made economic sense. The Tobin government, clearly reflecting public opinion in Newfoundland and Labrador, stubbornly refused to allow Inco to move Voisey's Bay ore offshore in anything less than a refined state. No basis for compromise appeared possible, and no resumption in negotiations was planned. As a result the Voisey's Bay Nickel Company was winding down its activities in the province, and Brian Tobin was rumoured to be considering a snap fall election with his handling of the Voisey's file as the primary issue. The spreading timelines continued to diverge.

■

This rupture of the close working relationship between Inco and the Newfoundland government was in itself remarkable. Only a year earlier the Tobin government had been Inco's staunchest ally, issuing precipitate (and illegal) permits for the infrastructure work objected to by the Innu and Inuit and opposing, in tandem with Inco, the Federal Court case brought by St. John's environmentalists to force a full environmental review of the smelter/refinery project. And yet this steady erosion of support for the Voisey's project under Inco's aegis, this alienation of Inco's allies on the ground in Newfoundland and Labrador, was another clearly defined trend.

How could all of this have happened? I believe the seeds of self-destruction of Inco, and of its Voisey's project, were clear in Mike Sopko's own words and attitudes at the Laurentian University commencement in 1995, which preceded Inco's involvement in Voisey's. Although he adjured the graduates to acknowledge and adapt to change in the modern world, he, and his company, failed to recognize and respond to most of the changes that lay outside the narrow ambit of their business world and therefore of their own experience.

In hindsight, perhaps no one should have been surprised that Inco would have such difficulty at Voisey's Bay. It had, after all, been more than 40 years since the company had developed a major ore body (Thompson, Manitoba) in North America, and much had changed in four decades. The rebirth of North American Aboriginal cultural and spirituality, which Ojibway elders Art and Eva Solomon had done so much to stimulate in Mike Sopko's own home town; the rising level of Aboriginal militancy and the trend toward recognition of Aboriginal title by the courts, which were clear even before Inco's first investment at Voisey's; the environmental movement; the history of economically and environmentally disastrous deals and policies in Newfoundland and Labrador; the advent of urban support groups, like the Voisey's Bay/Innu Rights Coalition, armed with the very sort of new technology and knowledge that Sopko seemed to be addressing in his commencement address: all of these "new ways of doing things" were addressed, in so many words, by Joan Kuyek right under Sopko's nose.

And yet rather than "keeping his mind open to new ideas," as Sopko himself urged the graduates to do, he chose to be embarrassed, to deny, and ultimately to ignore her words, and the truth of them. It was a pattern of behaviour that Sopko and his firm would

EPILOGUE

repeat over and over around Voisey's Bay. Again and again, in private and in public, well-intentioned and well-informed individuals, ranging from Sudbury lawyer Stephen O'Neill to Labrador Inuit leader William Barbour to Innu Nation presidents Peter Penashue and Katie Rich to Newfoundland premier Brian Tobin, would iterate and reiterate their demands/expectations/warnings about Voisey's Bay; and again and again Inco would choose to blithely ignore the clearly enunciated desires of their erstwhile partners.

In Guatemala or Indonesia or New Caledonia such loutish behaviour may still be acceptable, but in modern-day North America the mining industry's "stakeholders" are ignored only at the industry's own peril.

■

The breakdown in talks between Inco and Newfoundland over the smelter/refinery issue stole the headlines and stalled the Voisey's development yet again, but most of the other issues so critical to the project also remained unresolved. Here is a brief *tour d'horizon*, as of late summer 1998.

Inuit land claims—Negotiations that had been "fast-tracked" in the fall of 1996 with a March 31, 1997, deadline, and then broken off only to be resumed in the fall of 1997 and then proclaimed nearly competed by the *Globe & Mail* and Premier Tobin in November 1997, were still continuing. Lead LIA negotiator Toby Anderson told OkâlaKatigêt Society radio in August 1998 that negotiations had been adjourned until September, and that, while a number of issues remained outstanding, none appeared to be deal-breakers. Anderson speculated that an agreement-in-principle could be expected before Christmas 1998. It now appears possible, therefore, that the LIA may have a ratified comprehensive land claim agreement by, or shortly after, the turn of the century.

Innu land claims—Although the Innu land claims negotiations had begun long after those of the Inuit, the gap between the two appears to be closing fast, according to Christine Cleghorn. The Innu had presented their demands to the federal and provincial governments by the summer of 1998, and a senior-level bargaining session scheduled for July to discuss the Innu land position had been postponed until October 1998 at the request of the government negotiators. A best-case scenario is that the Innu will conclude a comprehensive land claims agreement in the early years of the 21st century. Even with Voisey's stalled the Tobin government has considerable motivation to reach an agreement with the Innu, because of its desire to develop the Lower Churchill Falls region for hydroelectric power.

Innu and Inuit IBA talks with Inco—Sporadic talks between the parties were still continuing, but Inco president Scott Hand signalled that they were no longer on the front burner, given that talks with the Newfoundland government had broken off. Resolution of IBA talks are absolutely crucial to production at Voisey's Bay, not least because the outcome will help determine the operating cost for the company. In all probability the IBA talks will not be concluded until after land claims are settled, which is what the Innu Nation and LIA wanted all along. Full IBAS, therefore, may not be in place with Inco until the middle of the first decade of the 21st century, if ever.

The Citizens' Mining Council court case—The original legal action, asking the Federal Court to force a full environmental assessment of the Argentia smelter/refinery, was filed in court in Toronto in September 1997. The court date was set for the following February in Vancouver, where the hearing did, in fact, begin. But it was adjourned after a few days because the presiding judge became ill. The case was rescheduled for a Toronto court in March, where the evidence was at last heard. The judge expected a decision by Easter, but as late August approached no judgment had been handed down. If the CMC wins its action it

EPILOGUE

could represent another serious delay to the Voisey's project, though Inco, of course, doesn't want to build the smelter/refinery any time soon in any case.

The overlapping Nunavik land claim—Remember the elders I encountered in the dining room in Nain? The Québec Inuit, or Nunavik people, continue to maintain their claim to most of Northern Labrador, including the Voisey's Bay area. And an early test of the legitimacy of the claim appears to have gone in their favour in the form of an August 1998 Federal Court ruling indicating that the federal government could not create a national park in the Torngat Mountains north of Nain without consulting the Québec, as well as the Labrador, Inuit. Lawyers for all sides were still evaluating the Federal Court decision in late August, but it appears the judgment will only further complicate the question of land tenure in Northern Labrador in general, and at Voisey's Bay in particular.

The environmental assessment process—This is about the only aspect of the Voisey's Bay development that appears to be more or less on schedule. Formal hearings were scheduled for the fall of 1998, and a report from the panel was expected by the spring of 1999.

In Sudbury, meanwhile, there was a limited sense of relief that Voisey's nickel would not be hitting the market any time soon, combined with admiration for how the Innu and Inuit and the government of Newfoundland had stood their ground against Inco. And, as the summer of 1998 waned, there were indications that the company was wearing out its welcome with the political leaders and public of Sudbury as well.

Both of the two Sudbury-area Liberal MPs said they would oppose any move to help Inco build a smelter in Argentia through tax breaks or other financial aid from the federal government. "It will be, basically, over my dead body," vowed Sudbury MP and long-time cabinet minister Diane Marleau. Meanwhile, Sudbury mayor Jim

Gordon said he understood Brian Tobin's position vis à vis a smelter and refinery in Newfoundland. "We have always said here in this community that we believe, first of all, the ore should be smelted here, we would like to see more added-value here, so therefore, why wouldn't any thinking Canadian—and in this case it happens to be the premier of Newfoundland—why would he say anything different than the people of Sudbury, or the mayor of Sudbury, for that matter?" Continuing the public backlash against Inco, a reader of the *Sudbury Star* opined, in a letter to the editor, that "Tobin should be given the Order of Canada for exposing the likes of Inco to the world. It perplexes me how a company can be so mismanaged and still be in business after 100 years. . . . Ontario politicians cater to Inco and let them destroy our economy and environment. Tobin has the right idea; show them up for what they are; if they go under someone else will take the money-making machine over, run it more efficiently and with more respect for the city and province. The only ones worse off when this happens will be Mike Sopko and cronies. The rest of the employees and citizens will have their lot improved and some respect."

But Sudburians also believed, grimly, that they were the ones who would pay the price for the Voisey's Bay fiasco. Inco appeared determined to slash its Sudbury workforce by 2,000 jobs by the year 2000, and the number of working mines was to be reduced to four—the fewest number of working mines at Inco's Sudbury property in the company's 96-year history.

Although company spokesmen denied it repeatedly, most Inco employees and Sudbury residents remained convinced that much of this bloodletting was to compensate for the time, energy, and treasure lost at Voisey's Bay.

■

And what of Voisey's Bay itself? Certainly with nickel prices at $1.85 U.S. per pound (down from $3.50 U.S. when Voisey Bay was

EPILOGUE

first purchased), and with world financial markets in turmoil, there is little need for the project that would have boosted annual world nickel production by 15 percent. Which is not to say that the Voisey's ore body won't be developed eventually. It certainly will, though perhaps not within the first decade of the 21st century, and not by Inco.

My guess is that Inco soon will acknowledge reality and write down the value of Voisey's by $1 to $2 billion, as Bay Street has long desired. This will represent a humiliating admission that the company paid far too much for the property in the first place, but it will be an important public confession for senior Inco management. I wouldn't be surprised to see Inco sell the Voisey's property in the end or to see all of once-mighty Inco gobbled up by a competitor. One thing is certain in all of this: Voisey's Bay has been absolutely ruinous to what had been the world's largest nickel producing company, and it will take years for Inco's balance sheet to recover from its flirtation with this premature bonanza, if it ever does.

Afterword and Acknowledgments

THIS BOOK had its genesis a year ago, when Jamie Swift, a long-time friend who is associated with the publisher Between the Lines, urged me to produce a manuscript for BTL on the Voisey's Bay development. I declined, arguing that it was impossible to tell a story properly, even one so compelling as the battle over Voisey's, when the outcome had yet to be determined. But Jamie and BTL refused to take no for an answer, and in the end, happily, their insistence wore down my resistance. I believe this book also owes a debt to another old friend, Mercedes Steedman, who may have put Jamie up to calling me in the first place.

I am also deeply indebted to *Northern Life* president (and friend) Mike Atkins, publisher John Thompson, and then-managing editor Carol Mulligan for their decision to send me to Labrador in the fall of 1996; it says a great deal about their feisty little newspaper that to this day I am the only Sudbury reporter to have made the trip to Voisey's Bay.

In Newfoundland and Labrador I must acknowledge the help of a host of new friends and acquaintances, including Leslie Bella,

AFTERWORD AND ACKNOWLEDGEMENTS

Fran Williams, Jim Brokenshire, Linda Whelan, Peter Evans, Laurie Heath, Paul Piggott, Christine Cleghorn, and Adrian Tanner.

This book could not have been written without the generous financial assistance of the Ontario Arts Council and Canada Council and the Sudbury Regional Credit Union; at the latter I must especially thank Karen Marconato, Mike Moore and Doug Yeo for their steadfast and ongoing support.

I am particularly indebted to Steve O'Neill, whose informal tutorials on Canadian law pertaining to Aboriginal matters opened up a whole new world for me that proved to have immense relevance for Voisey's Bay.

As ever, editorial support was indispensable, especially the invaluable critical input of managing editor Ruth Bradley-St-Cyr. Thanks also to Myna Wallin and Steve Osgoode, both interns at Between the Lines.

Finally, as always, I owe a special thanks to those closest to home who supported the writing of this book in ways financial, physical, and psychological—my wife Anne-Marie Mawhiney, my daughters Julia and Melanie Lowe, and my brother Doug Lowe. I love you all more than words can ever say. Thank you from the bottom of my heart.

Mick Lowe
Onwatin Lake, Ontario
November 1998

Appendix

Joan Kuyek's Commencement Speech

Sudbury, June 2, 1995

EVERY MORNING EARLY I go running with my dog Blue in the woods on the corner of Lasalle Boulevard and Frood Road. This spring the spicy smell of balsam poplar buds opening and the song of the white-throated sparrow greet us, and Blue leaps through the thick underbrush chasing rabbits and squirrels.

When I first moved here in 1970, the same place had been burnt to black rock by sulphuric acid from the open hearth roasters and I remember thinking that it looked like Mordor in the Lord of the Rings.

You can't live in this amazing city and not be moved by the changes in the landscape that scientists have made possible. But I also know that the real heroes and heroines of the re-greening are the nameless welfare recipients who were forced to work on land reclamation—a few at the cost of their lives and many at the cost of their backs—spreading lime and grass seed in the summers of the early 1980's. I also know that if the union wives and strikers hadn't made Inco the "most hated company in Canada" during the nine month 1978-79 Inco strike, the company would not have felt it necessary to improve its public image. And that if it weren't for the environmental movement, the people working in the company who wanted to develop ecologically sound policies would not have had enough leverage.

There is only one reality, but we all look at it through very different windows. Wisdom lies in learning to look through as many as possible. Some windows are harder to look through than others, they show us horror and pain, but unless we look through them and grasp what we see, we will not understand.

I believe we are put on this planet to be the best person we can be and that our life is an apprenticeship to get there. The values I have learned to name from my Anishnabi elders are kindness, sharing, honesty, courage, strength, humility and respect. I only hope that by the time I die, I have learned what it means to live by them.

Within us, we all have the ability to be kind, co-operative and honest; just as we all have the capacity to be mean, cruel and selfish. Some organizational forms bring out the good in us; some bring out the truly nasty parts. When we organize in hierarchies and give some people power over others, it enables those without power to abdicate responsibility for their actions. It allows us to do things in the course of our work that we would consider immoral and wrong in our home life.

Twenty years ago, I worked for Bell Telephone as a service representative, where part of my job was disconnecting telephones for non-payment of the bill. One day, I was faced with disconnecting a woman with no car, whose husband was unemployed, whose child was sick, and who lived out of town. They only owed three months regular service—no long distance. She pleaded that if the phone was disconnected her husband would never find work and she would be unable to get help for the child. The rules said I was to disconnect her. I broke down in the middle of the call, but my supervisor took over and did it without a qualm. She was "well trained."

On the other hand, when we work in equal relationships, we have to take responsibility for our own actions, and for one another.

At Better Beginnings where I work, the staff, who are all hired from the Donovan-Flour Mill neighbourhood, manage the project through team meetings and consultations. The board (or Council, as we call it) is elected from a membership of participants in our programs, including staff. We work in circles. People who visit the project say they have never seen anywhere before where everyone says they love their job. It is highly effective.

Organizations are systems of power relations: they can be egalitarian or hierarchical or combinations of the two. When a system provides a few with power over others, it corrupts the judgment of those who have that power and marginalizes those who don't. Our economy is one such system.

Many of us are completely intimidated by the assumptions and language of economics; we feel powerless to challenge or debate them. When we attempt to challenge these assumptions, we are told that we are not "realistic" and that we "don't understand good business practice." Economics has been elevated to the level of a religion in our society: it has its own priesthood, its own symbolic language and ritual observances (Christmas, Mother's Day, Easter) and it has a code of conduct for human affairs based on profit and the bottom line.

The central practice of modern economies is to turn relations and activities into "things" or "commodities" and to measure their value in currency (dollars, yen, rubles, etc.). Value comes to mean the amount of money that someone is willing or able to exchange for the commodity.

However, "Economy" comes from two Greek words meaning to "manage the household." Anyone in this room who has actually managed a household will see immediately how limiting the language of economics would be to describe it. There is no effective way to measure the quality of human relations, no way to measure work, food preparation, home repairs, composting, health, beauty or neighbourliness. What really matters in that home cannot be bought or sold. What has true value cannot be made into a commodity.

Under this system, wealth (from the Old English word for "health") has come to mean the accumulation of possessions, and the distribution of wealth, for those of us who don't own property or means of production, is determined by the "job"—selling one's labour for another's benefit.

In fact, there are many important kinds of work undertaken in a society: building a home or community, caring for children and for one another, and caring for the earth. These are done for "free" and in the language of economics have no value. All this work is part of the informal or shadow economy.

In the last fifty years, the world economy has devolved into the hands of a relatively few people who control the transnational corporations and financial markets. By 1994, 47 of the 100 largest economies in the world were not countries but corporations. This means that 138 countries had economies smaller than these giants.

In the past ten years corporate wealth has devolved into fewer and fewer hands. The number of billionaires doubled between 1987-92. By 1994, the richest fifth of the world's population consumed 83% of its wealth.

The demonstrators outside this convocation represent the Inco pensioners, a group of people who certainly reflect the contradictions of the economic system. They spent their working lives in an unsafe and dirty environment, employed by a company where, at one point, three-quarters of the workers left on disability pension, and where ten years back so many of them died right after retirement that the company couldn't even spend the interest earned on the pension plan.

Their precious time made stainless steel and a variety of useful products possible. It also created war materials. The hours they sold for wages created the wealth that pays executive staff incomes that range close to a million dollars and created the wealth that has been invested in Indonesia and Guatemala and that restricted the human rights of peoples in these countries. No wonder they are bitter.

At the same time, all corporations reward their CEOs in the same way: some pay even more. It is control over decision-making and the distribution of wealth that needs to be changed.

The consequences of this economic system are devastating: whole countries and economies are destroyed, war breaks out over scarce resources, the environment is raped and pillaged, topsoil blows away, refugees are driven from camp to camp, social security systems are wiped out, children starve and die of disease. These are the logical conclusions of an economics system based on the accumulation of wealth and power.

But other forces are also growing. Although the economic system appears monolithic, it must be remembered that its base is in fact quite vulnerable: the system depends on consenting human labour and on the steady flow of resources like electricity. When enough people who work in corporations stand up for things they believe in, the company itself will change direction. The "Acts of God" that are extraordinary natural occurrences like floods, earthquakes and forest fires work the same way. Competition for scarce markets and inputs and growing uncontrolled currency speculation make the whole system fragile.

Seeds of hope are sprouting everywhere around us, although they are ignored and undervalued because they are small and do not seek power over others or the earth through domination, force, or control. They are neighbourhood-based organizations, worker buyouts of companies, companies that become environmentally responsible, artists, musicians and story-tellers, earth-based spiritualities, co-operatives and micro-businesses that are sustainable, that respect human needs and that build self-reliance, organic agriculture and forestry, green technologies that heal the earth, and organizational structures where people work in safe, egalitarian ways.

APPENDIX

Hope lies, not in sweeping Utopian changes, but in the restructuring and rebuilding of our own community organizations to reflect the kinds of values that we believe in.

All over, people are refusing to be complicit in activities that pillage the earth and human relationships. We don't read a lot about it in the local press, but it goes on regardless. People resist mega-dams that destroy their homes in India, resist biotechnology and the patenting of human life forms, block logging trucks in Clayquot. The Innu claim their land in Nitassinan—where Voisey's Bay is located, and the Temaugama Anishnabi continue the struggle for Ndaki Menan. In our cities people work for more humane workplaces, bike paths, adequate welfare, affordable housing, enough daycare.

The work has been organizing me for some 30 years. It has never—until the past four years—been my paid job. It has been my vocation, my calling. And I have known I would do it no matter how I supported myself. Like me, most of you will not get paid to change the world in the ways we truly want.

Here is the work I have learned we can all do to build a world that will be safe for our grandchildren:

- Seek wisdom and tell the truth.
- Support technologies that heal the earth.
- Work with social, economic and political forums that foster sharing, equality, self-reliance and respect.
- Value those things that cannot be measured as much as those that are.
- Serve your own community with your money, your time, your work and your words.
- Be gentle with the earth and with one another.

Over 20 years ago, I had a dream. I was driving into Sudbury on Highway 17 in the midst of a thunderstorm. As I came near the smelter, I saw that it was standing on the backs of thousands of kneeling people: children, adults and youths, men and women, all the colours of human kind. They were sweating and groaning with effort. And then I saw the sun break through in the east and a rainbow appeared in the sky to my left. As I moved, the rainbow moved beside me until it collided with the super-stack. At that moment, the people under the smelter stood up. Grass and flowers and then trees grew up between them and around them. And the smelter, and the other enormous structures around it, crumbled into dust.

Thank you.

Annotated Bibliography

Brasch, Hans. *A Miner's Chronicle: Inco Ltd. and the Unions 1944-97.*
   Self-published by Brasch. Dowling, Ont, 1997
This book by a retired Sudbury miner contains, among other things, the first-ever listing of the names of all 673 men and one woman killed on the job at the Sudbury operations of Inco and its predecessor companies.

Clement, Wallace. *Hardrock Mining: Industrial Relations and Technological Changes at Inco.* McClelland and Stewart. Toronto, 1981
A sociologist, Clement mulled the meaning of tech change in the mining industry, and issued a prophetic warning as to its implications for workers. It says something that most of the "new" technology Clement studied is now either standard for the industry, or long since outmoded.

Culhane, Dara. *The Pleasure of the Crown: Anthropology, Law and First Nations.* Talonbooks. Vancouver, 1998
A monumental study of the issues and the lower court proceedings that led up to the landmark *Delgamuukw* decision.

Gunn, John M., editor. *Restoration and Recovery of an Industrial Region: Progress in Restoring the Smelter-Damaged Landscape near Sudbury, Canada.* Springer-Verlag. New York, 1995
This is a rare work—a collection of scientific and technical readings that is also highly readable. The last word in painstaking documentation of the decades of environmental damage caused by nickel smelting in the Sudbury district, and of the remarkable efforts to repair same. Lavishly illustrated.

Innu Nation and Mushuau Innu Band Council. *Gathering Voices: Finding Strength to Help Our Children.* Douglas & McIntyre. Vancouver/Toronto, 1995
Bilingual (English and *Innu-eimun*) account of the People's Inquiry established in Utshimassits-Davis Inlet in the wake of the house fire that killed six Innu children in February 1992. A handsome, and painfully candid, look at the social problems affecting the Innu, in their own words.

Innu Nation Task Force on Mining Activities. *Between a Rock and a Hard Place.* Innu Nation. Sheshatshiu, Nitassinan, 1996
Report of a Task Force established to determine the feelings of the Innu people about exploration and mining at Emish (Voisey's Bay). Contains a myriad of voices and opinions. About the only thing they have in common is their lucidity, forthrightness, and common sense.

Lowe, Mick. "Hacks, Flacks and Superstacks." Originally published in the August 1976 issue of *Content* magazine, anthologized in

ANNOTATED BIBLIOGRAPHY

The News, ed. Barrie Zwicker and Dick MacDonald. Deneau Publisher. Ottawa, 1982
Subtitled "Twenty five days on the Inco beat," a look at how Inco dominated Sudbury's news media in the spring of 1976.

_____ "Inco's Nickelmania." *Canadian Dimension*, Vol 12, No. 7, 1978
A detailed report on the Sudbury reaction to Inco's mass layoffs, "the largest industrial layoffs in a decade," in the fall of 1977.

_____ "Hard Rock Women." *The Financial Post Magazine*. January 1995
The experiences of women who broke the gender barrier, underground and on surface, at Inco's Sudbury operations.

_____ "Terms of Agreement." *The Financial Post Magazine*. April 1998
The lessons of Voisey's Bay and Raglan, and their implications for the future of the Canadian mining industry in the newly dawned age of Aboriginal land title.

North American Congress on Latin America. *Guatemala*. NACLA. Berkeley, 1974
Extensive background on Inco's ill-fated Exmibal nickel development.

Ross, Val. "The Arrogance of Inco." *Canadian Business Magazine*. May 1979
This indictment of Inco remains a classic of Canadian magazine reportage.

Royal Ontario Nickel Commission. *Report of the Royal Ontario Nickel Commission*. King's Printer. Toronto, 1917
An exhaustive *tour d'horizon* of the world nickel industry during the

First World War. Report led directly to the establishment of Inco's first nickel refinery in Canada a year later.

Seccaspina, Paul. *Structure and Change in the International Nickel Industry: Inco and the Transformation from Monopoly to Competition.* Unpublished Doctoral Thesis for University of Warwick. July 1997
The most comprehensive historical analysis ever written of Inco and its relations with governments, trade unions and environmentalists, from the company's founding in 1902 until 1993, just before the discovery of nickel at Voisey's Bay.

Slattery, Brian. "Understanding Aboriginal Rights." *The Canadian Bar Review.* Vol. 66, 1986
A landmark article on the rights of Canadian Aboriginal peoples that would influence an entire generation of Canadian jurists on the subject.

Solski, Mike and Jack Smaller. *Mine Mill: The History of the International Union of Mine, Mill and Smelter Workers in Canada since 1895.* Steel Rail Publishing. Ottawa, 1984
This partisan and lavishly illustrated history is as spunky as the union it so fondly recalls. Much information on Sudbury's Local 598, of which Solski was once president.

Swift, Jamie. *The Big Nickel: Inco at Home and Abroad.* Between the Lines. Toronto, 1977
First published more than 20 years ago, this anti-corporate history remains one of the best critical overviews of Inco.

Tester, Jim. *Son of a Working Man.* Laurentian Publishing Ltd. Sudbury, 1994
A collection of newspaper columns, originally published in *Northern Life*, Sudbury's community newspaper, by another former president of the Mine, Mill and Smelter Workers' Union Local 598.

Thompson, John F. and Norman Beasley. *For the Years to Come: A Story of International Nickel of Canada*. G. P. Putnam and Sons. New York, 1960
Still the primary corporate history of Inco, by a former president and CEO of the company. Thompson, Manitoba, was named in his honour.

Wadden, Marie. *Nitassinan: The Innu Struggle to Reclaim Their Homeland*. Douglas & McIntyre. Toronto, 1996
This winner of the Edna Staebler Award for Creative Non-Fiction is a first-rate introduction to the Innu Nation in general, and their struggle against low-level flying, in particular.

Internet Websites

As it has with so many of the public issues of our day, the Internet has become a useful, some might argue well-nigh indispensable, source of information about Voisey's Bay, from the standpoint of just about every stakeholder. Herewith, a list of the key websites, as of August 1998:

www.cancom.net/~franklia
Website of the Labrador Inuit Association. Not updated as regularly as the Innu Nation's site, but still a useful source of information on the Voisey's Bay development from the Labrador Inuit point of view. Includes the full text of the September 1997 ruling by the Newfoundland Court of Appeal that stymied Inco's construction program.

www.innu.ca
The trilingual (English, French and *Innu-eimun*) home page of the Innu Nation/Mamit Innuat. This is the *ne plus ultra* of Voisey's Bay-related websites. Very graphic and highly sophisticated, the site has received more than 30,000 "hits" since it was first posted in February

1996. A visitor can spend hours learning about Innu history, ethnography, and struggles against everything from low-level flying to Voisey's Bay itself. Given the Four Eagles Award by the Native American Indian Resources website, which described the Innu pages as "the best way to approach native culture or to learn about that of other tribes ... this site is a fine blend of both Kadizookaanag and Dibaajimowinan—traditional and carefully-weighed contemporary facts and narratives."

www.incoltd.com
Inco's website. Straightforward corporate site (nowhere near as flashy as the Innu's), containing recent speeches by company execs, news releases, financial statements, etc. Includes a special section on Voisey's Bay and a "Chronology of Nickel," a history of nickel from pre-historic times, told from the company's point of view. Still a useful reference, the history can be found under the "For Teachers Only" menu bar.

www.gov.nf.ca
Website of the government of Newfoundland and Labrador. Silent on Voisey's Bay *per se* of late, but has several pages on the Lower Churchill hydroelectric project.

www.techstocks.com
"The Silicon Investor" website, where putative "investors" gather in cyberspace to mull investment tips, gossip, and hard news. A Voisey's Bay "thread" was launched in July 1997, and it has provided highly entertaining reading ever since. It can be found under the Gold, Mining and Natural Resources department on the home page by scrolling down to Inco/Voisey's Bay, or you can go directly to www.techstocks.com/~wsapi/investor/Subject 16036.

www.makivik.org
Another trilingual website, in Inuktitut, English and French, the home pages of Makivik Corporation contain one of the great Canadian

Aboriginal business success stories of modern times. Makivik, the business arm of the Inuit of Northern Québec, has parlayed its government funding from the James Bay and Northern Québec Agreement into an impressive, and lucrative, investment portfolio, which includes the Raglan partnership, a string of airlines, including First Air ("Canada's third largest scheduled airline") and food processing companies.

www.falconbridge.com
Home page of Inco competitor, Raglan developer, and Makivik partner Falconbridge Ltd. An informative, easy-to-use corporate site from the company that may not be playing second fiddle to Inco much longer.

Glossary

Adjacency principle—the notion that communities adjacent to a mine that will endure the development's adverse environmental and social impacts have an inherent right to share directly in its benefits.

Comprehensive land claim—a block of land or lands, not necessarily contiguous, granted to a First Nation or Nations in return for the extinguishment of treaty or common law rights. Lands thus granted fall into one of three categories:

> **Category One Lands**—Lands on which Aboriginal beneficiaries hold surface and subsurface rights.
>
> **Category Two Lands**—Lands on which Aboriginal beneficiaries hold surface, but not subsurface rights.
>
> **Category Three Lands**—Land on which Aboriginal beneficiaries hold traditional hunting and fishing rights, but not surface or subsurface rights.

Environmental Impact Statement (EIS)—A developer's study into the environmental impacts of a proposed project, as mandated by the Canadian Environmental Assessment Act.

Impact and Benefits Agreement (IBA)—A legal contract between a First Nation and a private developer specifying the benefits that will accrue to the Aboriginal signatory in return for a guarantee of access and development rights on Aboriginal land.

James Bay and Northern Québec Agreement (JBNQA)—Signed on November 11, 1975 between the Québec and federal governments and the Cree of Northern Québec and the people of Nunavik, this first comprehensive land claim of the modern era paved the way for the massive James Bay hydroelectric development in Northern Québec.

Labrador Inuit Association (LIA)—The political organization, based in Nain, which represents the 5,000 or so Inuit of Northern Labrador.

Laterite ores—A type of nickel-bearing ore body, usually found in the tropics, that produces ferro-nickel, a primary ingredient in stainless steel alloys. Because most laterite deposits lie near the surface, they are mineable through relatively cheap open-pit methods, but the resulting ore is energy-intensive to smelt and lacks the valuable by-products associated with sulphide ore bodies. (See sulphide ores.) Inco's operations in Indonesia and New Caledonia and Falconbridge's properties in the Dominican Republic and New Caledonia are based on laterite ores.

Makivik Corporation—The business arm of the Nunavik people.

Nunavik people—The 8,500 Inuit, signatories of the James Bay and Northern Québec Agreement, of Northern Québec, or Nunavik.

Overlapping claim—A property claimed by more than one First Nation or party. Inco's Voisey's Bay property could be the subject of as many as five overlapping claims.

Sulphide ores—A type of nickel-bearing ore body that contains sulphur, along with various base and precious metals. Sulphide ores are usually, though not always, extracted through underground mining operations, which are more expensive than open-pit methods. Sudbury, Ontario, Thompson, Manitoba, Norilsk, in Northern Russia, and Voisey's Bay in Northern Labrador are among the world's greatest sulphide ore bodies.

GLOSSARY

The Whitehorse Mining Initiative (WMI)—The document, initiated by the Mining Association of Canada, which spells out a broad, overarching framework for mineral development in Canada in the 21st century.

Index

Allen, David, 92, 94, 95, 111-12
Anaktalak Bay, Nfld., 4, 36-37, 39, 42, 80-81
Anderson, Chesley, 77, 180
Anderson, Toby, 192
Anishnabi, 23
Archean Resources Ltd., 4, 35, 90
Argentia, Nfld., 61-62, 190
Ashcroft, Jim, 7, 20, 22
As It Happens, 22
Aylward, Kevin, 74, 98

Barbour, William, 58-59, 84, 192
Bardot, Brigitte, 39, 44
Berukoff, Walter, 27
Black, Conrad, 37
Bonin, Ray, 64
Bouchard, Lucien, 171, 176
Briggs, Rick, 128
Brokenshire, Jim, 52-53, 55, 61, 62, 91

Canadian Aboriginal Mining Association (CAMA), 57

Canadian Auto Workers, 127
Canadian Business, 12
Canadian Copper Company, 148
Canadian Environmental Defence Fund, 91
Canadian Helicopters Eastern, 37
Carson, Rachel, 96-97
Chislett, Al, 4-5, 35, 90, 96
Chrétien, Jean, 13, 64
Citizens' Mining Council of Newfoundland and Labrador, 157, 193
Clark, Herb, 73
Cleghorn, Christine, 68-70, 99, 165, 169-70, 174, 193
Company of Young Canadians, 9
Coniston, Ont., 36
Coon Come, Matthew, 127
Copper Cliff, Ont., 151-52
Corcoran, Terence, 95-96
Creighton Mine (Sudbury, Ont.), 35, 40
Curlook, Walter, 13

Daishowa Inc., 122-25
Davis Inlet, Nfld., 47, 80, 90
Delgamuukw, 133-38, 172
Diamond Fields Resources Ltd., 4, 25-26, 46, 58
Discovery Hill (Voisey's Bay, Nfld.), 36-37, 72
Dulles, Allen, 11, 112-13
Dulles, John Foster, 11, 112-13

Eastern Deeps, 35
Edison Building, 28
Eisenhower, Dwight, 11
Elliot Lake, 29-30
Emish, 4
Environmental Assessment Panel, 99
Evans, Peter, 147, 180-82

Falconbridge Ltd., 25-29, 90, 125
Ferguson, Charles, 152-53
Franklin, Sir John, (Canadian Coast Guard icebreaker), 37, 41-43
Friedland, Robert, 25-7, 29, 46, 58, 121
Friends of the Lubicon (FOL), 123-25
Frood Mine, 12

Gagnon, Johnny, 15, 17, 19
Gendron, Stewart, 34, 57-59, 60-61, 67, 69, 99
George River caribou herd, 3, 79
Gherson, Giles, 54
Gibbons, Rex, 51-55
Gill, Rick, 95
Gitskan First Nation, 133
Goose Bay, Nfld., 34, 47, 183
Gordon, Jim, 194
Gregoire, George, 165-66
Gregoire, Peter, 166

Green, Justice J.D., 96
Griffiths, Cathy, 158
Griffiths, Lesley, 74
Guatemala, 14, 112-13

Hamilton, Graeme, 37
Harris, Scott, 78, 199
Hopedale, 41
Hydro Québec, 60, 124

Inco, 16, 22, 27-29, 33, 37, 183
 buys 25 percent of DFR, 25
 buys remaining 75 percent, 29
 Canadian Copper Co. and Orford Copper Co., 148
 history of 10-15, 147-53
 in Indonesia and Guatemala, 111-13
 Ontario Division of, 7, 9, 16
 sintering plant, p. 15
Indonesia, 14, 111-13
Innes, Larry, 174
Innu Nation, 3-5, 25, 29, 46, 72, 77-78, 92, 114
Inuit of Labrador, 3-5, 25, 29, 46, 127-32
Iron Ore Co. of Canada, 163

James Bay, Québec, 29
James Bay Cree, 16
James Bay and Northern Quebec Agreement (JBNQA), 125
Jesuits, 16

Kanak, 115-17
Keevil, Norman, 27
Kinatuinamut Ilingajuk, 147, 180
Kingston, Ont., 9
Kuyek, Devlin, 21
Kuyek, Joan, 7-9; 17, 19-22; 191

INDEX

Labrador Inuit Association (LIA), 46, 58, 68, 72, 81, 92, 126, 128, 147, 180
Labrador Inuit Development Corp., 41, 80
Labrador Sea, 3, 36, 40
Lacey, Keith, 94
Laurentian University, 7, 14, 199
  board of governors of, 21, 23
  history and physical setting of,.15
Lethbridge, Kirk, 183
Lower Churchill hydro project, 171, 176, 193
Lubicon Crees, 122-25
  relations with Inco, 16, 22

McCreedy East mine (Sudbury, Ont.), 28
Makivik Corporation, 46, 126
Manitoba, University of, 9
Mansbridge, Peter, 11
Marleau, Diane, 194
Marshall, Justice W.W., 96
Massacre Island, Labrador., 47
Memorial University of Newfoundland, 39
Merkuratsuk, Jacko, 40, 44
Mine, Mill and Smelter Workers' Union, 11, 16, 127, 150-51
Miramar Mining Corp., 27
Mowat, Farley, 181
Mulligan, Carol, 30
Mulroney, Brian, 163
Munday, Tony, 94
Mushuau Band of the Innu Nation, 78-79; 82-83, 162

Nain, Labrador, 4, 30, 33-34, 36-37, 39, 40, 41-42, 44-45, 147
Napier, Bill, 162-164

Natuashish (Sango Bay), 184
New Caledonia, 113
Newfoundland Supreme Court decision (Sept. 22, 1997), 96-99
Nitassinan, 4, 60, 78-79, 82, 170
Northern Life, 30, 43
Northwest Territories, 36
Nunavik, 127, 194
Nungak, Zebedee, 46

Oka, 60
OkâlaKatigêt Communications Society, 47, 83-84, 179
O'Neill, Steve, 25, 119-22, 132, 134, 192
O'Rourke, Mike, 42, 49
Orr, James, 148
Ovoid, The, 35-36, 158

Pamak, Richard, 39-44
Parker, Ralph, 16, 19
Paul, Ross, 7, 20-22
Penashue, Peter, 57-60, 192
Penny, Jackie, 161
Piwas, Akat, v
Prince Charles, 13
Pugsley, Thomas, 128

Rae, Bob, 13
Raglan, 28, 90, 125
Ramsey, Lake, 16, 21
Rich, Katie, 74-75, 77-79, 83-85, 87-89, 96, 170-77, 185, 192
Rich, Monique, 86-87, 162, 164, 166-68
Roberts, Ed, 173
Ross, Val, 12
Royal Ontario Nickel Commission, 149
Russia, 13

Sailian, Janet, 18
Sango Bay, 184
Saturday Night, 13
St. John's, 34
Séguin, Homer, 114
Sheridan, Patrick, 27
Simpson, Jeffrey, 125
Sheshatshiu, 47, 78, 81, 83, 86
Slattery, Brian, 120
Solomon, Art, 23, 191
Solomon, Eva, 23, 191
Sopko, Mike, 58, 175
  announces 25% purchase of DFR, 25
  commencement speech, 189, 191
  investors conference in London, U.K., 94-95
  shareholders' meeting (1997), 67-68, 72
  Steve O'Neill and, 119-120
Southam News, 37
Soviet Union, 13
Stanley, Gail, 34
Steele, Justice G.L., 96
Stewart, Christine, 74
Sudbury, University of, 16
Suzuki, David, 123

Tanner, Adrian, 78
Teck Corp., 27
Ten Mile Bay, Labrador, 41
Thomas, Kevin, 124
Thompson, John, 30
Thompson, Robert, 149
Tobin, Brian, 48, 93, 171-72, 176, 190, 195

Tompkins, Harry, 12

United Steelworkers of America, 11-12, 16, 127, 181

Vallée, Emil, 114
Verbiski, Chris, 4-5, 35, 90, 96
Victor Mine (Sudbury, Ont.), 28
Voisey's Bay, Nfld.
  description and discovery of, 3-5
  first Inco announcement concerning, 20
  impact on Sudbury 27-30
  latitude of, 36
  orebody description, 35
Voiseys' Bay/Innu Rights Coalition, 70, 121, 123-124, 191
Voisey's Bay Nickel Co. (VBNC), 34, 45, 48, 57, 61, 73, 84, 90, 93, 158-59, 161, 174, 179, 182

Wadden, Marie, 82
Walkom, Thomas, 93-94
Wallace, Jamie, 7, 20, 22
Watson, Patrick, 123
Western Extension, 35
Wet'suwet'en First Nation, 133
Whitehorse Mining Initiative (WMI), 128
Williams, Fran, 47, 132, 180-82
Winslow, Donna, 115-17
Winsor, Hugh, 172

Yukon Territory, 36